OUR ECOLOGICAL FOOTPRINT

REDUCING HUMAN IMPACT
ON THE EARTH

Mathis Wackernagel and William E. Rees

Illustrated by Phil Testemale

NEW SOCIETY PUBLISHERS

Canadian Cataloguing in Publication Data

Wackernagel, Mathis, 1962 -
  Our ecological footprint

  (The new catalyst bioregional series; 9)
  Includes bibliographical references.
  ISBN 1-55092-250-5 (bound) -- ISBN 1-55092-251-3 (pbk.)
  1. Sustainable development. 2. Man -- Influence on nature.
3. Human ecology. 4. Economic development -- Environmental aspects.
I. Rees, William E. II. Title. III. Series.
HC79.E5W32 1995   333.7   C95-910838-6

Cover design by David Lester, from an illustration by Phil Testemale.

Printed in Canada on partially recycled acid-free paper using soy-based ink by Transcontinental Printing.
Eleventh printing

Inquiries regarding requests to reprint all or part of *Our Ecological Footprint: Reducing Human Impact on the Earth* should be addressed to New Society Publishers at the address below.

Canada ISBN: 1-55092-251-3 (Paperback)
USA ISBN: 0-86571-312-X (Paperback)

To order directly from the publishers, please add $4.00 shipping to the price of the first copy, and $1.00 for each additional copy (plus GST in Canada). Send check or money order to:

New Society Publishers,
P.O. Box 189, Gabriola Island, B.C. Canada V0R 1X0

New Society Publishers aims to publish books for fundamental social change through nonviolent action. We focus especially on sustainable living, progressive leadership, and educational and parenting resources. Our full list of books can be browsed on the world wide web at:
http://www.newsociety.com

Our Ecological Footprint is number 9 in *The New Catalyst*'s Bioregional Series of books. For others in the series, please write: New Society Publishers, Canada.

NEW SOCIETY PUBLISHERS
Gabriola Island, BC

The New Catalyst Bioregional Series

The *New Catalyst* Bioregional Series was begun in 1990, the start of what some were calling "the turnaround decade" in recognition of the warning that humankind had ten years to turn around its present course, or risk such permanent damage to planet Earth that human life would likely become unviable. Unwilling to throw in the towel, *The New Catalyst*'s editorial collective took up the challenge of presenting, in new form, ideas and experiences that might radically influence the future.

As a tabloid, *The New Catalyst* magazine had been published quarterly since 1985. From the beginning, an important aim was to act as a catalyst among the diverse strands of the alternative movement—to break through the overly sharp dividing lines between environmentalists and aboriginal nations; peace activists and permaculturalists; feminists, food co-ops, city-reinhabitants and back-to-the-landers—to promote healthy dialogue among all these tendencies working for progressive change, for a new world. The emerging bioregional movement was itself a catalyst and umbrella for these groups, and so *The New Catalyst* became a bioregional journal for the northwest, consciously attempting to draw together the local efforts of people engaged in both resistance and renewal from as far apart as northern British Columbia, the Great Lakes, and the Ozark mountains, as well as the broader, more global thinking of key people from elsewhere in North America and around the world.

To broaden its readership, *The New Catalyst* changed format, the tabloid reorganized to include primarily material of regional importance, and distributed free, and the more enduring articles of relevance to a wider, continental audience now published twice yearly in *The New Catalyst*'s Bioregional Series. Through this new medium, we hope to encourage on-going dialogue among overlapping networks of interest, to solidify our common ground, expand horizons, and provoke deeper analysis of our collective predicament as well as a sharing of those practical, local initiatives that are the cutting edge of more widespread change.

The Bioregional Series aims to inspire and stimulate the building of new, ecologically sustainable cultures and communities in their myriad facets through presenting a broad spectrum of concerns ranging from how we view the world and act within it, through efforts at restoring damaged ecosystems or greening the cities, to the raising of a new and hopeful generation. It is designed not for those content with merely saving what's left, but for those forward-looking folk with abundant energy for life, upon whom the future of Earth depends.

Other volumes in the series include *Turtle Talk: Voices For A Sustainable Future*, (No.1), *Green Business: Hope or Hoax?* (No.2), *Putting Power In Its Place: Create Community Control!* (No.3), *Living With The Land: Communities Restoring The Earth* (No.4), *Circles of Strength: Community Alternatives to Alienation* (No.5) and *Boundaries of Home: Mapping for Local Empowerment* (No.6), and *Futures by Design: The Practice of Ecological Planning* (Nos. 7 & 8 — a double volume).

TABLE OF CONTENTS

Acknowledgments ix
Preface x

INTRODUCTION 1
Why Worry About Sustainability? 1
What We Hope to Achieve 3
A Matter of Perspective 4

1. ECOLOGICAL FOOTPRINTS FOR BEGINNERS 7
Obvious but Profound: We Depend on Nature 7
What *is* an Ecological Footprint? 9
So What? The Global Context 13
Dr. Footnote Explains 16
The Power of Science 17
The Wisdom of the Marketplace 18
The Doctrine of Free Trade 20
The Uncertain Future 22
The Technological Fix 23
The Mantra of Optimism 25
The Growth of Limits 26
Planning for a Sustainable Future 28

2. FOOTPRINTS AND SUSTAINABILITY 31
The Sustainability Debate: A Simple Concept Leads to Conflicting
Strategies 31
The sustainability challenge 31
*Strong sustainability: the ecological bottom-line condition
for sustainability* 36
The Brundtland Commission's proposed response 39
The Ecological Footprint: A Tool for Planning Toward Sustainability . 40
Measuring progress toward sustainability: the dos and don'ts 40
Learning from ecology: revisiting human carrying capacity 48
Turning carrying capacity on its head: human Ecological Footprints . . . 51
How Ecological Footprint analyses can help advance sustainability . . . 55

3. FUN WITH FOOTPRINTS: METHODS & REAL-WORLD APPLICATIONS 61

Making the Ecological Footprint Idea Work 61
Calculation Procedure 63
   Consumption categories 66
   Land and land-use categories 69
   The consumption – land-use matrix 77
The Footprint in Action: Adapting the Calculation Procedure
to Specific Applications 79
   1) How big is the Ecological Footprint of the average North American? . 80
   2) How large is the Vancouver regional Footprint? 86
   *3) A global comparison of Footprint sizes — could everybody on Earth
      today enjoy North Americans' current ecological standard of living?* 88
   4) Footprinting Great Britain 91
   *5) European examples: The Ecological Footprints of the Netherlands
      and the Trier Region of Germany* 93
   6) A Regional Analysis from Australia 96
   7) What does ecological dependence mean for trade? 96
   8) Is a person's Ecological Footprint related to income? 100
   9) Housing choice affects our Footprints 103
   *10) How much ecologically productive land supports commuting
      by bicycle, bus and car?* 106
   11) Did you know that tomatoes leave Footprints? 108
   12) The Ecological Footprint of bridges 109
   13) Learning about sustainability in schools and in the outdoors . . . 111
   14) State of the Environment Reporting 113
   15) Interpreting sustainability: The ecological "Rorschach test" 113
   16) Calculate your own Footprint 116
   17) Eco-labelling: Is your product sustainable? 118

4. THE SEARCH FOR SUSTAINABILITY STRATEGIES . . . 125

Questioning Conventional Strategies 127
The Process of Developing Sustainability 133
   *The two sustainability poles: ecological stability and human
      quality of life* 133
   Win-win solutions 135
   The cycle of change in decision-making 137
   Three uphill battles to achieve sustainability 139
Sketching a Vision for a Sustainable Society 140

5. AVOIDING OVERSHOOT: A SUMMARY **149**
  Creating Public Awareness 150
  Developing Sustainability — Locally and Globally 154
  GLOSSARY 158

List of Boxes

  BOX 2.1: Sustainability and Sustainable Development:
    Some Clarification 33
  BOX 2.2: On Natural Capital 35
  BOX 2.3: Strong or Weak Sustainability? 37
  BOX 2.4: The Entropy Law and the Economy/Ecology Conundrum . 43
  BOX 2.5: A Brief History of the Human Carrying Capacity Concept . . 48
  BOX 3.1: The Human Footprint in the Sea 64
  BOX 3.2: Data Sources for Ecological Footprint Analyses 70
  BOX 3.3: Some Examples: Translating Consumption into Land Areas . 80
  BOX 3.4: Assessing the Footprint of the Netherlands 95
  BOX 3.5: The Ecological Deficits of Industrialized Countries 97
  BOX 3.6: Calculating the Footprint of India 98
  BOX 3.7: Determining the Footprint of Commuting 106
  BOX 3.8: The Ecological Footprint of a Newspaper 119
  BOX 4.1: Will Efficiency Gains Save Resources? 128

List of Tables

  Table 3.1: The eight main land and land-use categories for
    footprint assessments 68
  Table 3.2: Productivity of various energy sources 72
  Table 3.3: The consumption land-use matrix for the average
    Canadian 82
  Table 3.4: Comparing people's average consumption in the U.S.,
    Canada, India and the World 85

ACKNOWLEDGMENTS

We would like to thank our colleagues from the University of British Columbia Task Force on Healthy and Sustainable Communities — Peter Boothroyd, Mike Carr, Lawrence Green, Clyde Hertzman, Judy Lynam, Sharon Manson-Singer, Janette McIntosh, Aleck Ostry and Robert Woollard — for their support and encouragement. Our research on Ecological Footprint analysis, including Mathis Wackernagel's PhD studies, was supported in part by a Canadian Tri-Council EcoResearch grant to U.B.C. in which the Task Force shared.

Preface

Some years ago, I read of a species of tiny woodland wasp that lives on mushrooms. It seems that when a wandering female wasp chances upon the right kind of mushroom in the forest, she deposits her eggs within it. Almost immediately, the eggs hatch and the tiny grubs begin literally to eat themselves out of house and home. The little maggots grow rapidly, but soon something very odd happens. The eggs in the larvaes' own ovaries hatch while still inside their immature mothers. This second generation of parthenogenic grubs quickly consumes its parents from within, then breaks out of the empty shells to continue feeding on the mushroom. This seemingly gruesome process may repeat itself for another generation. It doesn't take long before the entire mushroom is over-filled by squirming maggots and fouled by their bodily wastes. The exploding population of juvenile wasps consumes virtually its entire habitat which is the signal for the largest and most mature of the larvae to pupate. The few individuals that manage to emerge as mature adults then abandon their mouldering birthplace, flying off to begin the whole process over again.

We wrote this book in the belief that the bizarre life-cycle of the mushroom wasps may offer a lesson to humankind. The tiny wasps' weird reproductive strategy has apparently evolved under extreme competitive pressure. Good mushrooms — like good planets — are hard to find. Natural selection therefore favored those individual wasps and reproductive traits that were most successful in appropriating the available supply of essential resources (the mushroom) before the competition had arrived or became established.

No doubt human beings also have a competitive side and both natural and sociocultural selection have historically favored those individuals and cultures that have been most successful in commandeering resources and exploiting the bounty of nature. There is also plenty of archeological and historic evidence that, like the over-crowded mushroom, many whole cultures have collapsed from the weight of their own success. Human societies as temporally and spatially far-flung as the Mesopotamians, Mayans, and Easter Islanders likely came to ruin by expanding beyond the capacity of their environments to sustain them. Like the forest wasps, they depleted their local habitats. Humanity as a whole survived, however, because there were always other figurative "mushrooms" elsewhere on Earth capable of supporting people.

Today, of course, humankind has become a global culture, one increasingly driven by a philosophy of competitive expansionism, one which is subduing and consuming the Earth. The problem is that, unlike the wasp, even the fattest and richest among us have no means to abandon the withered hulk of our habitat once consumed and there is no evidence yet of other Earth-like "mushrooms" in our galactic forest.

The good news is that — also unlike the wasp — humans are gifted by the potential for self-awareness and intelligent choice, and *knowing our circumstances is an invitation to change.*

The first step toward reducing our ecological impact is to recognize that the "environmental crisis" is less an environmental and technical problem than it is a behavioral and social one. It can therefore be resolved only with the help of behavioral and social solutions. On a finite planet, at human carrying capacity, a society driven mainly by selfish individualism has all the potential for sustainability of a collection of angry scorpions in a bottle. Certainly human beings are competitive organisms but they are also cooperative social beings. Indeed, it is no small irony (but one that seems to have escaped many policy advisors today) that some of the most economically and competitively successful societies have been the most internally cooperative — those with the greatest stocks of cultural and social capital.

Our primary objective with this book is to make the case that we humans have no choice but to reduce our "Ecological Footprint." We hope that it also conveys our essential confidence in the resourcefulness of the human spirit. People have great untapped potential to meet this greatest of challenges to our collective security. As William Catton stated in his 1980 classic, *Overshoot*: "If, having overshot carrying capacity, we cannot avoid crash, perhaps with ecological understanding of its real causes we can remain human in circumstances that could otherwise tempt us to turn beastly." Indeed, we believe that confronting together the reality of ecological overshoot will force us to discover and exercise those special qualities that distinguish humans from other sentient species, to become truly human. In this sense, global ecological change may well represent our last great opportunity to prove that there really *is* intelligent life on Earth.

William Rees
Gabriola Island
Summer 1995

Introduction

Humans are facing an unprecedented challenge: there is wide agreement that the Earth's ecosystems cannot sustain current levels of economic activity and material consumption, let alone increased levels. At the same time, economic activity on the globe as measured by Gross World Product is growing at four percent a year, which corresponds to a doubling time of about 18 years.[1] One factor driving this expansion is the growth of the world's population: in 1950, there were 2.5 billion people, while today there are 5.8 billion. There may well be 10 billion people on Earth before the middle of the next century. Even more ecologically significant is the rise in *per capita* energy and material consumption which, in the last 40 years, has soared faster than the human population. An irresistible economy seems to be on a collision course with an immovable ecosphere.

Why Worry About Sustainability?

The conventional approach to development has been highly successful at expanding economic activity and economic growth remains at the forefront of most nations' political agendas. The long-term goal is to integrate local and national economies into one global economy with unrestricted trade and capital flows. This will greatly boost industrial production and likely further increase resource consumption. However, the weaknesses in the conventional model are more and more apparent. For example, increasing economic production has neither levelled income differences, made the "haves" noticeably happier, nor satisfied the basic needs of the world's poorest one billion people. While 20 percent of the world's population enjoys unprecedented material well-being, at least another 20 percent remain in conditions of absolute poverty. In fact, the top 20 percent of income earners take home over 60 times more than the poorest 20 percent, and this gap has doubled over the last 30 years.[2] Conventional economic development has been challenged for this glaring social inequity since its inception with the Bretton Woods agreements after the Second World War.

Today, in the face of ecological constraints, the criticism is even more severe. Current rates of resource harvesting and waste generation deplete nature faster than it can regenerate. Stanford University biologist Peter Vitousek and his

1

colleagues calculated in 1986 that human activities were by then already "appropriating," directly or indirectly, 40 percent of the products of terrestrial photosynthesis — in effect, humanity was channelling through its economy 40 percent of nature's land-based biological production — and more recent work suggests that human exploitation of the continental shelves is approaching similar levels. If the human use of other natural functions of nature is included, such as waste absorption by the land and water, and the protection from harmful ultraviolet radiation (by the stratospheric ozone layer), it is not hard to imagine that human activities may be using the world beyond long-term capacity.

The accelerating resource consumption that has supported the rapid economic growth and the rising material standards of industrialized countries in recent decades has, at the same time, degraded the forests, soil, water, air and biological diversity of the planet. As the world becomes ecologically over-

Why Worry? As the world becomes ecologically overloaded, conventional economic development becomes self-destructive and impoverishing — and puts human survival at risk (after Horst Haitzinger).

loaded, conventional economic development actually becomes self-destructive and impoverishing. Many scholars believe that continuing on this historical path might even put our very survival at risk. Certainly, there is little to indicate that current sustainability initiatives will be effective at reversing global ecological deterioration. Indeed, pressure on both ecological integrity and social health is mounting. More effective sustainability initiatives are required, including tools to stimulate a wider public involvement, evaluate strategies and monitor progress.

What We Hope to Achieve

This book describes a planning tool that can help to translate sustainability concerns into public action: we call it "Ecological Footprint" analysis. The Ecological Footprint concept is simple, yet potentially comprehensive: it accounts for the flows of energy and matter to and from any defined economy and converts these into the corresponding land/water area required from nature to support these flows. This technique is both analytical and educational. It not only assesses the sustainability of current human activities, but is also effective in building public awareness and assisting decision-making. The Ecological Footprint is not about "how bad things are." It is about humanity's continuing dependence on nature and what we can do to secure Earth's capacity to support a humane existence for all in the future. Understanding our ecological constraints will make our sustainability strategies more effective and livable. Ecological Footprint analysis should help us to choose wisely, which we think is preferable to having nature impose a choice of its own.

Thus, to the extent that Ecological Footprint analysis reflects biophysical reality it is *good news* for a better and more secure future. The *bad news* is the conventional dream that the human enterprise can be expanded forever on a finite planet. This expansionist vision might sound attractive, but it is bound to fail in its current form. This failure would be very painful. It would hurt the poor first, the rich a little later, and all the way along destroy many of our fellow species.

The Ecological Footprint approach acknowledges that humanity is facing difficult challenges, makes them apparent, and directs action toward sustainable living. Admittedly, acknowledging the darker side of the human condition is sometimes painful — avoidance is sweet temptation. However, this book takes the position that denial today leads to greater pain tomorrow. We believe that the first step toward a more sustainable world is to accept ecological reality and the socioeconomic challenges it implies. Any "business as usual" strategy that perpetuates today's destructive lifestyles would be a disservice to our children.

A Matter of Perspective

To develop a way of living that is fulfilling *and* sustainable within nature, we need to rethink our relationships with each other and with the rest of nature. This book tries to stimulate such thinking. There are, of course, many books with a similar purpose but we hope this one is a little different.

To begin with, most writers on the subject — even the good ones — treat the "environment" as something *out there*, separate and detached from people and their works. This is, in fact, a fair reflection of our prevailing cultural ethic. Judging from our actions and our language, modern humans generally tend to see society as more or less independent of nature. Thus, when economic activity causes unexpected damage to some environmental value, we call it a "negative externality," emphasizing the environment's place on the periphery of modern consciousness. Little wonder that conventional approaches to development treat the environment as a kind of backdrop to human affairs! The environment may be aesthetically pleasing, but it is expendable if economic push comes to shove. The loss of environmental value is still seen as an unfortunate but mostly necessary "trade-off" against economic growth. The well-worn old saw "you can't stop progress" catches the prevailing ethic pretty well.

This book starts from a different premise. We argue that the human enterprise cannot be separated from the natural world even in our minds because there is no such separation in nature. In terms of energy and material flows, there is simply no *"out there"* — the human economy is a fully dependent sub-system of the ecosphere. This means that we should study humanity's role in nature in much the same way we would study that of any other large consumer organism. The fact is that through the economic production-con-sumption-pollution cycle, humankind has become a major — and often the dominant — species in virtually every significant ecosystem on the planet.

The premise that *human society is a subsystem of the ecosphere,* that human beings are embedded in nature, is so simple that it is generally overlooked or dismissed as too obvious to be relevant. However, taking this "obvious" insight seriously leads to some profound conclusions. The policy implications of this ecological reality run much deeper than pressing for improved pollution control and better environmental protection, both of which maintain the myth of separation. If humans are part of nature's fabric, the "environment" is no mere scenic backdrop but becomes the play itself. The ecosphere is where we live, humanity is dependent on nature, not the reverse. Sustainability requires that our emphasis shift from "managing resources" to managing *ourselves,* that we learn to live as part of nature. Economics at last becomes human ecology.

This book shows that we can develop more sustainable lifestyles. We propose tools and frameworks for understanding the challenges, evaluating strategies and monitoring progress, and provide examples of how these

The Ecological Footprint is a measure of the "load" imposed by a given population on nature. It represents the land area necessary to sustain current levels of resource consumption and waste discharge by that population.

strategies work. Achieving sustainability will require much thought and sweat, but changing the world can also be pretty exciting.

We have tried to appeal to a diverse audience, and hope to offer something to every level of interest. Chapter one describes and illustrates the Ecological Footprint concept. Chapter two links it to the sustainability debate. The next chapter explains the procedure for Footprint calculations and discusses 17 applications. Finally, we conclude the book with a discussion of sustainability strategies and a summary of what we have learned.

Notes

1. The Gross World Product rose from $3.8 trillion in 1950 to $19.3 trillion in 1993 (measured in 1987 US dollars). Worldwatch Institute, *Vital Signs 1994* (NY: W. W.

Norton, 1994).
2. United Nations Development Program (UNDP), *Human Development Report* (NY: Oxford University Press, 1992, 1994).

1

ECOLOGICAL FOOTPRINTS FOR BEGINNERS

Many of us live in cities where we easily forget that nature works in closed loops. We go to the store to buy food with money from the bank machine and, later, get rid of the waste either by depositing it in the back alley or flushing it down the toilet. Big city life breaks natural material cycles and provides little sense of our intimate connection with nature.

Obvious but Profound: We Depend on Nature

Despite this estrangement, we are not just *connected* to nature — we *are* nature. As we eat, drink and breath, we constantly exchange energy and matter with our environment. The human body is continuously wearing out and rebuilding itself — in fact, we replace almost all the molecules in our bodies about once a year. The atoms of which we are made have already been part of many other living beings. Particles of us once roamed about in a dinosaur, and some of us may well carry an atom of Caesar or Cleopatra.

Nature provides us with a steady supply of the basic requirements for life. We need energy for heat and mobility, wood for housing and paper products, and nutritious food and clean water for healthy living. Through photosynthesis green plants convert sunlight, carbon dioxide (CO_2), nutrients and water into chemical energy (such as fruit and vegetables), and all the food chains that support animal life — including our own — are based on this plant material. Nature also absorbs our wastes and provides life-support services such as climate stability and protection from ultraviolet radiation. Finally, the sheer exuberance and beauty of nature is a source of joy and spiritual inspiration. Figure 1.1 shows how very tightly human life is interwoven with nature, a connection we often forget or ignore. Since most of us spend our lives in cities and consume goods imported from all over the world, we tend to experience nature merely as a collection of commodities or a place for recreation, rather than the very source of our lives and well-being.

If we are to live sustainably, we must ensure that we use the essential products and processes of nature no more quickly than they can be renewed, and that we discharge wastes no more quickly than they can be absorbed. Even

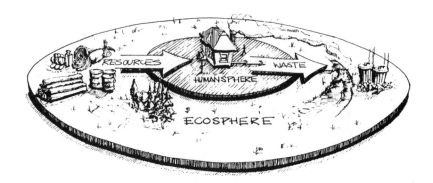

Figure 1.1: We are part of nature. Nature supplies material requirements for life, absorbs our wastes, and provides life-support services such as climate stabilization, all of which make Earth hospitable for people.

today, however, accelerating deforestation and soil erosion, fisheries collapse and species extinction, the accumulation of greenhouse gases and ozone depletion all tell us our current demands on nature are compromising humanity's future well-being. In spite of these trends, society operates as if nature were an expendable part of our economy. For example, agriculture, forestry and fisheries are considered to be mere extractive sectors of the economy, and since such primary activities contribute relatively little to the Gross National Product (GNP) of most industrialized countries, they are not considered to be particularly important. This perspective forgets that nature's products are indispensable to human well-being, however "insignificant" their dollar contribution to the country's GNP might be. Similarly, some people reduce the economy-ecology connection to pollution that directly threatens the health of people (e.g., urban air pollution). No doubt, this is an important problem but the emphasis on human health betrays a narrow ecological understanding. The economy's growing demands on nature endanger the planet's ability to support life on a much more fundamental level. Over-harvesting and waste generation not only reduce future productivity, but can lead to ecosystems collapse. So far, this phenomenon has been confined to the local or regional level (desertification in the African Sahel and the loss of North Atlantic groundfish stocks being recent examples). However, increasing evidence of global change is clear warning that human activity may now be undermining global life-support systems. The prospect of significant climate change, with its potential threat to food production and the safety of coastal settlements, should *in itself* be sufficient to force society to adopt a less cavalier attitude toward "the environment" that sustains us (to say nothing of 30 million other species).

What *is* an Ecological Footprint?

Ecological footprint analysis is an accounting tool that enables us to estimate the resource consumption and waste assimilation requirements of a defined human population or economy in terms of a corresponding productive land area. Typical questions we can ask with this tool include: how dependent is our study population on resource imports from "elsewhere" and on the waste assimilation capacity of the global commons?, and will nature's productivity be adequate to satisfy the rising material expectations of a growing human population into the next century? William Rees has been teaching the basic concept to planning students for 20 years and it has been developed further since 1990 by Mathis Wackernagel and other students working with Bill on UBC's Healthy and Sustainable Communities Task Force.

To introduce the thinking behind Ecological Footprint analysis, let's explore how our society perceives that pinnacle of human achievement, "the city." Ask for a definition, and most people will talk about a concentrated population or an area dominated by buildings, streets and other human-made artifacts (this is the architect's "built environment"); some will refer to the city as a political entity with a defined boundary containing the area over which the municipal government has jurisdiction; still others may see the city mainly as a concentration of cultural, social and educational facilities that would simply not be possible in a smaller settlement; and, finally, the economically-minded see the city as a node of intense exchange among individuals and firms and as the engine of production and economic growth.

No question, cities are among the most spectacular achievements of human civilization. In every country cities serve as the social, cultural, communications and commercial centers of national life. But something fundamental is missing from the popular perception of the city, something that has so long been taken for granted it has simply slipped from consciousness.

We can get at this missing element by performing a mental experiment based on two simple questions designed to force our thinking beyond conventional limits. First, imagine what would happen to any modern city or urban region — Vancouver, Philadelphia or London — as defined by its political boundaries, the area of built-up land, or the concentration of socioeconomic activities, if it were enclosed in a glass or plastic hemisphere that let in light but prevented material things of any kind from entering or leaving — like the "Biosphere II" project in Arizona (Figure 1.2). The health and integrity of the entire human system so contained would depend entirely on whatever was initially trapped within the hemisphere. It is obvious to most people that such a city would cease to function and its inhabitants would perish within a few days. The population and the economy contained by the capsule would have been cut off from vital resources and essential waste sinks, leaving it both to starve and to suffocate at the same time! In other words, the ecosystems contained within our imaginary human terrarium would have insufficient "carrying capacity" to support the ecological load imposed by the contained human population. This mental model of a glass hemisphere reminds us rather abruptly of humankind's continuing ecological vulnerability.

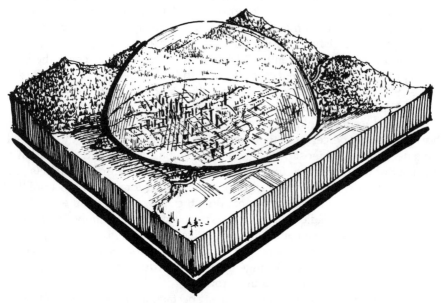

Figure 1.2: Living in a Terrarium.
How big would the glass hemisphere need to be so that the city under it could sustain itself exclusively on the ecosystems contained?

The second question pushes us to contemplate this hidden reality in more concrete terms. Let's assume that our experimental city is surrounded by a diverse landscape in which cropland and pasture, forests and watersheds — all the different ecologically productive land-types — are represented in proportion to their actual abundance on the Earth, and that adequate fossil energy is available to support current levels of consumption using prevailing technology. Let's also assume our imaginary glass enclosure is elastically expandable. The question now becomes: how large would the hemisphere have to become before the city at its center could sustain itself indefinitely and exclusively on the land and water ecosystems and the energy resources contained within the capsule? In other words, what is the total area of terrestrial ecosystem types needed continuously to support all the social and economic activities carried out by the people of our city as they go about their daily activities? Keep in mind that land with its ecosystems is needed to produce resources, to assimilate wastes, and to perform various invisible life-support functions. Keep in mind too, that for simplicity's sake, the question as posed does not include the ecologically productive land area needed to support other species independent of any service they may provide to humans.

For any set of specified circumstances — the present example assumes current population, prevailing material standards, existing technologies, etc. — it should be possible to produce a reasonable estimate of the land/water area required by the city concerned to sustain itself. By definition, the total ecosystem area that is essential to the continued existence of the city is its *de facto* Ecological Footprint on the Earth. It should be obvious that the Ecological Footprint of a city will be proportional to both population and *per capita* material consumption. Our estimates show for modern industrial cities the area involved is orders of magnitude larger than the area physically occupied by the city. Clearly, too, the Ecological Footprint includes all land required by the defined population wherever on Earth that land is located. Modern cities and whole countries survive on ecological goods and services appropriated from natural flows or acquired through commercial trade from all over the world. The Ecological Footprint therefore also represents the corresponding population's total "appropriated carrying capacity."

By revealing how much land is required to support any specified lifestyle indefinitely, the Ecological Footprint concept demonstrates the continuing material dependence of human beings on nature. For example, Table 3.3 (pages 82-83) shows the Ecological Footprint of an average Canadian, i.e., the amount of land required from nature to support a typical individual's present consumption. This adds up to almost 4.3 hectares, or a 207 metre square. This is roughly comparable to the area of three city blocks. The column on the left shows various consumption categories and the headings across the top show corresponding land-use categories.

"Energy" land as used in the Table means the area of carbon sink land required to absorb the carbon dioxide released by *per capita* fossil fuel consumption (coal, oil and natural gas) assuming atmospheric stability as a goal. Alternatively, this entry could be calculated according to the area of cropland necessary to produce a contemporary biological fuel such as ethanol to substitute for fossil fuel. This alternative produces even higher energy land requirements. "Degraded Land" means land that is no longer available for nature's production because it has been paved over or used for buildings. Examples of the resources in "Services" are the fuel needed to heat hospitals, or the paper and electricity used to produce a bank statement.

To use Table 3.3 to find out how much agricultural land is required to produce food for the average Canadian, for example, you would read across the "Food" row to the "Crop" and "Pasture" columns. The table shows that, on average, 0.95 hectares of garden, cropland and pasture is needed for a typical Canadian. Note that none of the entries in the table is a fixed, necessary, or recommended land area. They are simply our estimates of the 1990s ecological demands of typical Canadians. The Ecological Footprints of individuals and whole economies will vary depending on income, prices, personal and prevailing social values as they affect consumer behavior, and technological sophistication — e.g., the energy and material content of goods and services.

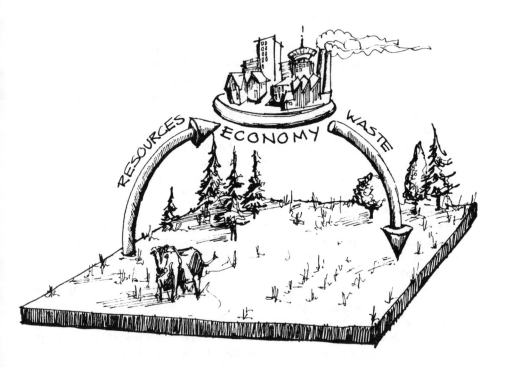

Figure 1.3: What is an Ecological Footprint?
Think of an economy as having an "industrial metabolism." In this respect it is similar to a cow in its pasture. The economy needs to "eat" resources, and eventually, all this intake becomes waste and has to leave the organism — the economy — again. So the question becomes: how big a pasture is necessary to support that economy — to produce all its feed and absorb all its waste? Alternatively, how much land would be necessary to support a defined economy sustainably at its current material standard of living?

Figure 1.4: Your Footprint. The average North American Footprint measures 4 to 5 hectares or is comparable to three-plus city blocks.

So What? — The Global Context

Our economy caters to growing demands that compete for dwindling supplies of life's basics. The Ecological Footprint of any population can be used to measure its current consumption and projected requirements against available ecological supply and point out likely shortfalls. In this way, it can assist society in assessing the choices we need to make about our demands on nature. To put this into perspective, the ecologically productive land "available" to each person on Earth has decreased steadily over the last century (Figure 1.5). Today, there are only 1.5 hectares of such land for each person, including wilderness areas that probably shouldn't be used for any other purpose. In contrast, the land area "appropriated" by residents of richer countries has steadily increased. The present Ecological Footprint of a typical North American (4–5 ha) represents three times his/her fair share of the Earth's bounty. Indeed, if everyone on Earth lived like the average Canadian or American, we would need at least three such planets to live sustainably (Figure 1.6). Of course, if the world population continues to grow as anticipated, there will be 10 billion people by 2040, for each of whom there will be less than 0.9 hectares of ecologically productive land, assuming there is no further soil degradation.

Such numbers become particularly telling when used to compare selected geographic regions with the land they actually "consume." For example, in Chapter 3 we estimate the Ecological Footprint for the Lower Fraser Valley, east of Vancouver to Hope, B.C. This valley bottom has 1.8 million inhabitants for a population density of 4.5 people per hectare. In short, the area is far smaller than needed to supply the ecological resources used by its population. If the average person in this basin needs the output of 4.3 hectares (Table 3.3), then

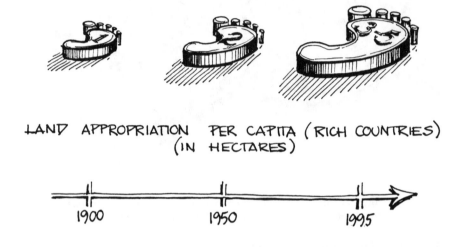

LAND APPROPRIATION PER CAPITA (RICH COUNTRIES)
(IN HECTARES)

1900   1950   1995

ECO-PRODUCTIVE LAND AREA AVAILABLE PER CAPITA (WORLD)
(IN HECTARES)

5.6   .25 ARABLE

Figure 1.5: Our Ecological Footprints Keep Growing While Our per capita "Earth-shares" Continue to Shrink. Since the beginning of this century, the available ecologically productive land has decreased from over five hectares to less than 1.5 hectares per person in 1995. At the same time, the average North American's Footprint has grown to over 4 hectares. These opposing trends are in fundamental conflict: the ecological demands of average citizens in rich countries exceed *per capita* supply by a factor of three. This means that the Earth could not support even today's population of 5.8 billion sustainably at North American material standards.

Figure 1.6: Wanted: Two (Phantom) Planets. If everybody lived like today's North Americans, it would take at least two additional planet Earths to produce the resources, absorb the wastes, and otherwise maintain life-support. Unfortunately, good planets are hard to find...

the Lower Fraser Valley depends on an area 19 times larger than that contained within its boundaries for food, forestry products, carbon dioxide assimilation and energy (Figure 3.5). Similarly, Holland has a population of 15 million people, or 4.4 people per hectare, and although Dutch people consume less than North Americans on average, they still require about 15 times the available land within their own country for food, forest products and energy use (Figure 3.8, Box 3.4). In other words, the ecosystems that actually support typical industrial regions lie invisibly far beyond their political or geographic boundaries.

A world upon which everyone imposed an over-sized Ecological Footprint would not be sustainable — the Ecological Footprint of humanity as a whole must be smaller than the ecologically productive portion of the planet's surface. This means that if every region or country were to emulate the economic example of the Lower Fraser Basin or the Netherlands, using existing technology, we would all be at risk from global ecological collapse.

The notion that the current lifestyle of industrialized countries cannot be extended safely to everyone on Earth will be disturbing to some. However, simply ignoring this possibility by blindly perpetuating conventional approaches to economic development invites both eco-catastrophe and subsequent geopolitical chaos. To recognize that not everybody can live like people do in industrialized countries today is not to argue that the poor should remain poor. It is to say that there must be adjustments all round and that, if our ecological analyses are correct, continuing on the current development path will actually hit the less fortunate hardest. Blind belief in the expansionists' cornucopian dream does not make it come true — rather it side-tracks us from learning to live within the means of nature and ultimately becomes ecologically and socially destructive.

Dr. Footnote Explains

Various critics have raised well-reasoned objection to aspects of the Ecological Footprint concept. In this section, sustainability counsellor Dr. Footnote addresses some of the issues they have raised.

THE POWER OF SCIENCE:

Analytical Scientist:
The Ecological Foot-
print is much too
pretentious. For
example, in spite of
years of detailed
and systematic re-
search, we still do
not know exactly
how single organisms
work (be they bac-
teria or blue whales),
and we know even
less about how they
interact. We scien-
tists work with
models, but they are

crude simplifications — and we can never prove them right.
The best we can do is prove them wrong. As good scientists
we must acknowledge our enormous ignorance of nature.
We need to be humble. So, how can you claim that the complex
interactions between
people and nature
can be reduced to a
matter of hectares?

Dr. Footnote: You're
right. The Ecological
Footprint doesn't tell
the whole story.
However, while
many people strive
toward the absolute
truth, a more relevant
question is whether
the knowledge we
use is compatible
with the phenomena
that we observe.
Know-ledge needs

to be appropriate to the task. For example, Newton's mechanical laws were good enough to fly us to the moon, in spite of their shortcomings in light of Einsteinian relativity. Not knowing something with certainty should not deter us from taking action or counter-action. Let's avoid paralysis by analysis, but rather err on the safe side. We must advocate precaution where potential danger looms — even if we do not know the exact nature of the hazard.

The Ecological Footprint model may be simple — like any ecological model, it does not represent all possible inter-actions. However, it estimates the minimum land area necessary to provide the basic energy and material flows required by the economy. We don't look at pollution beyond carbon dioxide. If anything, therefore, our current Ecological Footprint calculations underestimate humanity's draw on nature.

Even so, our calculations show that people have overshot global carrying capacity and that some people contribute significantly more to that overshoot than others. It is questionable, of course, whether humanity's Ecological Footprint should even approach the size of the Earth. Only a smaller Footprint provides any ecological resilience in the face of global change. In any case, today's ecological overshoot can only be temporary, and comes at a high cost to the future.

In short, we may not know exactly how nature works, but by using fundamental laws and known relationships we can calculate useful (under)estimates of human demands. They may not be precise enough for managing nature, but they do provide challenging guidelines for managing ourselves in an ecologically and socially more responsible way.

THE WISDOM OF THE MARKETPLACE:

Business person:
The trends are clear. Global income is rising faster than human population. Illiteracy is declining. Agri-cultural production has increased because it responds to growing demand. Life on the planet is better than ever. If we have environmental problems it is only because property rights are poorly defined or prices do not reflect the true costs. Once we get the prices right, the "Invisible Hand" will take care of those problems. Prices are the most effective way to tell people what to do and what not to

do and government interference should be kept to a minimum. Society's needs will then be met as people pursue their own individual interests.

Dr. Footnote: You're right, to a point. When nature's goods and services are underpriced they become over-used and abused, and the "Invisible Hand" that is supposed to automatically balance the market becomes the destabilizing "Invisible Elbow." Thus, adjusting prices through depletion taxes and pollution charges, for example, can be effective in reducing activities that are ecologically destructive. However, the Invisible Hand may often depend on the Ecological Footprint to work its magic. Ecological Footprint analysis may help us to assess the true social costs of growth because it makes visible many impacts to which traditional monetary analysis is usually blind. But let's be realistic, the "free market" will not solve all our problems. Not everything of value can (or should) be privatized and not all nature's services can even be quantified, let alone priced.

(What's the market price of a stable and predictable climate? How much ozone layer is enough?) The fact is that many decisions about people, resources and the ecosphere will continue to rely on partial scientific information and political judgment. Even such economic incentives as resource depletion taxes and tradable pollution rights require government intervention in the economy.

By the way, there is nothing inconsistent between your global economic trends and Ecological Footprint analysis. Higher incomes mean greater access to resources and bigger Ecological Footprints for the privileged minority. However, superabundance today does not guarantee even adequacy tomorrow. Much of our present "income" is derived from the liquidation of natural capital. Our Footprints are expanding even as the land upon which we stand shrinks beneath us.

THE DOCTRINE OF FREE TRADE:

Pilot: It seems to me that the Ecological Footprint questions the value of trade. I don't want to live in the Middle Ages! Trade is beneficial to everyone. For example, in North America, we cannot grow coffee and bananas, while coffee and banana exporters may not be able to build computers or grow wheat. Also, it is more economically efficient if we

produce the products where it is ecologically most efficient. For example, is it not stupid to grow winter tomatoes in heated greenhouses in Canada rather than import them from California or Mexico?

BUT PREVAILING TERMS OF TRADE UNDERMINE SUSTAINABILITY

Dr. Footnote:
Ecological Footprint analysis is not against trade per se. However, it examines trade through an eco-logical lens and reveals its environmental consequences. When economists talk about trade balances they refer only to money flows, not ecological flows. The fact is that some areas constantly give up ecological productivity, while others con-tinuously draw on it. For example, Hong Kong, Switzerland and Japan, which have positive dollar trade balances, provide little ecological productivity to the world, while importing a great deal from other places to maintain their high levels of consumption. Unfortunately, not everybody can be a net im-porter of ecological goods and services. On the global scale, for every importer there must be an exporter. This means that even though most developing nations are trying to follow the development of places like Japan, Hong Kong or Switzerland, it is physically impossible for all of them to succeed.

Expanding world trade leads to increased global resource flows, which stimulates total economic production and accelerates the depletion of the planet's natural assets — and there are other problems. People who live on ecological goods imported from afar (and on "common-pool" ecological functions such as climate control, which are shared by every-one) are spatially and psychologically disconnected from the resources that sustain them. They lose any direct incentive to conserve their own local resources and have no hand in the management of the distant sources of supply. In fact, they may remain blissfully unaware of both the ecological and social effects of prevailing terms of trade. Modern intensive production methods not only accelerate the depletion and contamination of field and forest, but the economic benefits of the increased productivity are inequitably distributed, particularly in low-income countries. Those who need the income may actually be displaced from the land to make way for export crops while the profits flow mainly to the already well-off. In short, in a world where the global economy is already pressing ecological limits and poverty still stalks a billion people, we don't need "free trade," but terms of trade that encourage the rehabilitation of natural capital and direct the benefits of export activities to those who need them most.

THE UNCERTAIN FUTURE:

Fortune Teller:
Ecological Footprint analysts claim to see the future. But predictions and extrapolations are always way off. The only thing we know about the future is that it is likely to be different from what we think it will be. Even I have difficulty seeing into the future with my crystal ball...

THE FUTURE SEEMS FOGGY AND UNCERTAIN — HOW CAN THE ECOLOGICAL FOOTPRINT PREDICT ANYTHING ?

Dr. Footnote:
Ecological Footprint analysis is not a predictive tool. It is an "ecological camera" that takes a snapshot of our current demands on nature. Extrapolation to the anticipated human population and resource flows in 2040 does suggest there are serious biophysical barriers on our current development path, but the numbers do not predict how things will turn out. Rather, they measure the "sustainability" gap that society must somehow close to ensure a stable future. In short, Ecological Footprint analysis can show how much we have to reduce our consumption, improve our technology, or change our behavior

FOOTPRINT ANALYSIS DOES NOT PREDICT, IT SIMPLY TAKES AN ECOLOGICAL SNAPSHOT

to achieve sustainability. It can also reveal with graphic clarity the chronic material inequity that persists between affluent and low-income countries today. Most important, Ecological Footprint analysis suggests some of the ways society can begin the shift toward sustainability and which of these measures provide the greatest leverage. To reiterate, this tool is not a telescope into the future, but a way to visualize the consequences of current trends and to assess alternative "what if" scenarios on the road to sustainability.

THE TECHNOLOGICAL FIX:

Robot: For hundreds of years people have worried that we would run out of land or resources. But no: the technological revolution has increased the abundance and lowered the prices of goods and services. Thanks to technology, a single farmer produces more than 200 farmers did 200 years ago. Thanks to

TECHNOLOGY CAN FIX IT

technology, millions of people in North America live more comfortably, are healthier, feel more secure and eat better than even kings and queens could dream of a few hundred years ago.

Who could have anticipated the computer revolution? Who can anticipate the future benefits of genetic engineering? For the last two hundred years, technology has successfully met the challenges of growth. Once people are faced with a problem, they will come up with a solution. Our greatest resource is the human mind, and the potential for innovation is unlimited. Just think about recent advances in medicine, transportation and communications. Why shouldn't we be able to fix any problem in the future?

POSSIBLY, BUT CERTAINLY NOT TODAY'S TECHNOLOGY AND NOT WITH CURRENT ECONOMIC INCENTIVES

Dr. Footnote:
Ecological Footprint analysis does not question the importance of technological innovation. In fact, technology will play a major role in making society more sustainable.
If we really want to build a global economy five to 10 times the size of today's (as suggested by the Brundtland report), then we need technology that makes us five to 10 times more resource-efficient. Some analysts already refer to this as the "factor-10" economy (see chapter 4).

Clearly, improved technologies are essential. Even simple things like solar water heaters or better insulation in our houses can reduce our Footprint without compromising our material standards of living. However, keep in mind that many technological innovations have not reduced our use of resources, but only substituted capital — resources and machines — for labor. For example, while modern agriculture produces more output per farmer than traditional agriculture, it requires much more energy, materials and water per unit of crop produced (as the tomato example shows in Chapter 3). Also, in present circumstances, gains in technological efficiency often encourage increased consumption — more efficient cars are more economical and are consequently used more frequently by more people. Indeed, in spite of efficiency gains, most industrial countries' total energy consumption has increased in recent years. In this context, the Ecological Footprint can be an important measuring rod of progress toward sustainability. Can new technology increase or reduce society's demand on nature? It depends; if new technology is to reduce our Ecological Footprint, it must be accompanied by policy measures to ensure that efficiency gains are not redirected to alternative forms of consumption.

THE MANTRA OF OPTIMISM:

Optimist: Ecological Footprint analysis is depressing. It paints a bleak picture of the future. People like you seem to have an affinity for apocalyptic visions. Such visions have existed all through human history, but they have never come true. Why do you not look on the bright side of life? Stop to smell the roses — let's have a good time!

CHEER-UP BUDDY

Dr. Footnote: Acknowledging that nature has a finite capacity is not pessimistic, just realistic. It makes room for wise decisions. To ignore these basic constraints would jeopardize future well-being. Ecological Footprint analysis starts from the premise that humanity must live within global

carrying capacity. It also maintains that if we choose wisely it might even be possible to increase our quality of life. Our concern is that the way we now live on the planet is self-destructive. The Footprint is a tool that facilitates learning about ecological constraints and developing a sustainable lifestyle. The earlier humanity starts to act upon the new challenges, the easier it will be.

THE GROWTH OF LIMITS:

Energy Producer:

Energy is the driving force of the human enterprise. If we have enough energy, we can do anything we like: clean up the environment, irrigate deserts, build fast transport-ation networks, power highly productive greenhouses — you name it! Today's ecological scarcity is only temporary. It won't be long before we develop

unlimited energy sources. Fusion energy is promising and we have hardly tapped into the potential for conventional fission power. And, imagine the potential if we could use all the tidal wave or solar energy that goes to waste today!

<u>Dr. Footnote</u>: Some people do hope that humanity will be able to harness unlimited energy supplies. In fact, we already are endowed with a huge energy source: the sun beams 175,000 terawatts to our planet, compared to just 10 terawatts of commercial energy, mainly fossil fuel, used by the human economy. However, imagine the impact

PERHAPS, BUT USING WHAT WE HAVE WISELY WOULD GIVE US MORE THAN ENOUGH

of an unlimited energy supply, if not used wisely or with restraint. We've run down much of the planet with just 10 terawatts! Unlimited cheap energy could simply expand human activities further, depleting other natural capital stocks until we run into some new — and probably more severe — limiting factor. It may not be energy resources, but the waste assimilation capacity of our planet, that becomes most limiting. For example, while we used to be concerned about running out of fossil fuel, scientists now realize that CO_2 sinks are even scarcer (they're already filled to overflowing).

Of course, used with due caution, technology can help to overcome ecological scarcity. Indeed, moving toward a solar economy may be the most promising strategy for reducing our Ecological Footprint. Solar energy, with all its necessary equipment, will be more expensive, and we will use it more wisely. However, with a solar economy we should be able to secure a higher future quality of life.

Planning for a Sustainable Future

The Ecological Footprint is a tool to help us plan for sustainability. It not only addresses such global concerns as ecological deterioration and material inequity, it also links these concerns to individual and institutional decision-making. Further refinement is necessary to develop the tool's full potential for planning practitioners' everyday decisions. However, it has already been applied in over 20 different situations, including those presented as examples in this book. In these applications, which range from environmental outdoor education for children to policy and project assessments for municipalities, Ecological Footprint analysis is already helping to frame sustainability issues and solutions in Canada and several other countries.

Ecological deterioration and social injustice can be reversed — there are

Figure 1.7: Paths We Can Choose.
What kind of future would you like and how can we get there?

thousands of conceptual tools and inspiring ideas about how to plan for a safer and more secure world. The Ecological Footprint is one of these tools. It helps us to understand both our present situation and the implications of policy choices.

Ecological Footprint analysis helps to put things in the larger perspective. To return to a previous image, we interpreted the Footprint of a city as the total area that would have to be enclosed with the city under a glass capsule to sustain the consumption patterns of the people in that city. Even without actual data, this mental image illustrates an important reality: as a result of high population densities, the rapid rise in *per capita* energy and material consumption, and the growing dependence on trade (all of which are facilitated by technology), *the ecological locations of human settlements no longer coincide with their geographic locations.* Modern cities and industrial regions are dependent for survival and growth on a vast and increasingly global hinterland of ecologically productive landscapes.

There is a small irony here — many science fiction writers have also evoked the image of a domed city, but in science fiction the device is usually needed to isolate and protect the human habitat from a hostile external environment. By contrast, our capsule experiment emphasizes that, without free access to the "environment," it is the isolated human habitat that becomes hostile to human life!

Thinking about such an encapsulated city forces us to consider not only all the ways in which we remain dependent on nature, but also on all the ways we can reduce humankind's negative impact on the systems that sustain us. For example, assume for a moment that *your* city or community is confined within a human terrarium as described above. That is, the hemisphere containing your city is just adequate to sustain the present population at prevailing material standards. Now ask yourself what the planning process and land-use bylaws might look like in the urban capsule. What sort of decision-making process would there be and who would be involved? What "trade-offs" and development costs that we currently ignore suddenly become very important? What criteria might be used to decide between private interests and the common good? To make this really interesting and more concrete, compare the desired planning process and legal regime with those currently in use in your community. Why are they different? Do these differences really make sense when we consider that the ecosphere is nothing but one big capsule containing the entire human family? The following chapters take off from here to explain how the Ecological Footprint concept contributes to building a sustainable society.

Notes

1. Michael Jacobs, *The Green Economy: Environment, Sustainable Development and the Politics of the Future* (Vancouver: U.B.C. Press, 1993) — originally published by Pluto Press in 1991.

2

FOOTPRINTS
AND SUSTAINABILITY

Confusion about the meaning of sustainability and why it matters has slowed progress toward achieving it. This confusion is not completely innocent but sometimes reflects the deliberate blurring of issues and conflicts of interest, as well as genuine fears. In this chapter, we try to untangle the confusion; we argue that sustainability is a simple concept — at least conceptually — and suggest that pondering the implications of the Ecological Footprint model helps us to understand at least the ecological requirements for a sustainable society.

The Sustainability Debate: A Simple Concept Leads to Conflicting Strategies

The sustainability challenge

Ever Since Rachel Carson's *Silent Spring* appeared in 1962, a burgeoning literature has substantiated the concern that the ecosphere, our life-support system, is being eroded at an accelerating pace. The list of threats to the life-support system in which we are embedded is overwhelming: deserts are encroaching on ecologically productive areas at the rate of 6 million hectares per year; deforestation claims over 17 million hectares per year; soil oxidation and erosion exceeds soil formation by 26 billion tonnes per year; fisheries are collapsing; the draw-down and pollution of ground water accelerates in many places of the world; as many as 17,000 species disappear every year; despite corrective action, stratospheric ozone continues to erode; industrial society has increased atmospheric carbon dioxide by 28 percent. All these trends are the result of either over-exploitation (excessive consumption) or excessive waste generation.[1] Since everything we consume eventually joins the waste stream, it is a convenient shorthand to say that the energy and material "throughput" of the human economy is beyond safe limits.

At the same time, many people are unable to meet even their basic requirements. As noted in the Introduction, 20 percent of the human population enjoys unprecedented wealth, including the bulk of the people in the "North." However, 20 percent earning less than 1.4 percent of the global income endures conditions of constant malnutrition. This segregation, accentuated by gender and ethnicity, goes beyond income. The fact that in 1990, just 3.5 percent of the world's cabinet ministers were women, and that 93 countries were without female ministers at all, is a visible symptom of a much deeper social inequality.[2]

Concerned people have advocated a more responsible and equitable use of the ecosphere throughout the 20th century, but it was not until 1987 that *Our Common Future*, the report of the United Nations World Commission on Environment and Development (or Brundtland Commission), popularized the idea of "sustainable development." The destructive social and ecological effects of the prevailing approach to "development" had finally become a serious item on the political agenda.

The starting point for the Brundtland Commission's work was their acknowledgment that the future of humanity is threatened. *Our Common Future* opened by declaring:

> The Earth is one but the world is not. We all depend on one biosphere for sustaining our lives. Yet each community, each country, strives for survival and prosperity with little regard for its impacts on others. Some consume the Earth's resources at a rate that would leave little for future generations. Others, many more in number, consume far too little and live with the prospects of hunger, squalor, disease, and early death.[3]

To confront the challenges of over-consumption on the one hand and grinding poverty on the other, the Commission called for *sustainable development*, defined as "...development that meets the needs of the present without compromising the ability of future generations to meet their own needs." In other words, the Commission recognized that the conventional economic imperative to maximize economic production must now be constrained — or perhaps we should say augmented — by both an ecological imperative to protect the ecosphere and a social imperative to minimize human suffering, today and in the future. For the first time, environment and equity became explicit factors in the development equation. Sustainable development therefore depends both on reducing ecological destruction (mainly by limiting the material and energy throughput of the human economy) and on improving the material quality of life of the world's poor (by freeing up the ecological space needed for further growth in developing countries and ensuring that the benefits flow where they are most needed).

Starting from the Brundtland definition, we argue that, conceptually, sustainability is a *simple* concept: it means living in material comfort and peacefully with each other within the means of nature. Despite this seeming simplicity, however, there is no general agreement on the policy implications of the concept (see Box 2.1). Some people are unconvinced there is a sustainability crisis at all, and others are frightened of the implications of

BOX 2.1: Sustainability and Sustainable Development: Some Clarification[4]

The need for humanity *to live equitably within the means of nature* is the underlying message of most definitions of sustainable development beginning with the Brundtland Commission's widely accepted call to "...[meet] the needs of the present without compromising the ability of future generations to meet their own needs." However, despite the widespread acknowledgement of the ecological and social symptoms of the problem, interpretations of sustainable development and its implications are contradictory, even within the Brundtland Commission's report.

One reason for conflicting interpretations of the fundamental sustainability message is obvious — the term "sustainable development" is itself treacherously ambiguous. Many people identify more with the "sustainable" part and hear a call for ecological and social transformation, a world of environmental stability and social justice. Others identify more with "development" and interpret it to mean more sensitive growth, a reformed version of the *status quo.* Sharachchandra Lélé writes that the various interpretations of sustainable development are caused not by poor understanding, but rather by ideological differences and reluctance of many to acknowledge the implications of the underlying message. The deliberate vagueness of the concept, even as defined by Brundtland, is a reflection of power politics and political bargaining, not a manifestation of insurmountable intellectual difficulty. Michael Redclift comments that "...unless we are *prepared to interrogate our assumptions* about both development and the environment and *give political effect* to the conclusions we reach, the reality of unsustainable development will remain...."

As suggested above, some of the confusion around "sustainable development" is rooted in general failure to distinguish between true development and mere growth. Economist Herman Daly clarifies the difference by defining "growth" as an increase in size through material accretion while referring to "development" as the realization of fuller and greater potential. In short, growth means getting bigger while development means getting better. For Daly, then, "sustainable development" is progressive social betterment without growing beyond ecological carrying capacity. Indeed, he regards "sustainable growth" as a nonsensical self-contradiction. Developing sustainability may actually require a *reduction* in aggregate economic throughput, while enabling the poor to consume *more.*

There are other ambiguities hidden in "sustainable development." It could refer to: a) the necessary conditions to live sustainably (a goal or state of being); b) the sociopolitical means of achieving the goal (a planning process); or, c) particular strategies to solve present problems (piecemeal solutions). Failure to clarify how the term is being used in a specific context can lead to fruitless misunderstanding. To some ears, the term *"developing sustainability"* is less ambiguous and is to be preferred over *"sustainable development."*

TECHNOLOGICALLY ACCELERATED
RESOURCE USE

SUSTAINABLE
RESOURCE USE

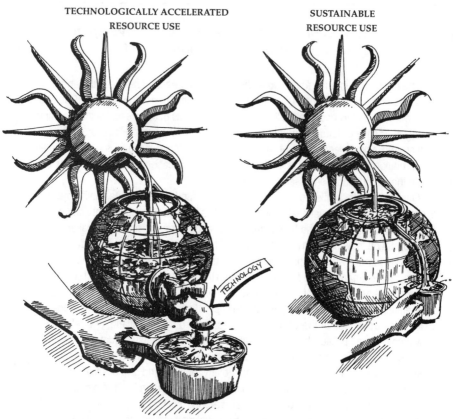

Figure 2.1: Sustainable Use: The Water Bucket Analogy.
Imagine a bucket being filled with water at a fixed rate. The water in the bucket is a capital stock that can be drawn upon only as rapidly as the bucket is being refilled. This balanced withdrawal rate is a form of sustainable income. Similarly, nature is a "bucket" that is continuously replenished by the sun: photosynthesis produces plant matter, the basis for all biological capital and most other life, and climatic, hydrological, and other biophysical cycles are solar powered too. Sustainability implies that nature's capital should be used no more rapidly than it can be replenished (right). However, trade and technology have enabled human-kind progressively to exploit nature far beyond sustainable levels so that present consumption exceeds natural income (the "interest" on our capital). This leaves the next generation with depleted capital and less productive potential even as the population and material expectations increase (left).

acknowledging that there is.

Of course, if environmental scientists are correct (and we believe they are) the consequences of *not* acknowledging material constraints on the economy

BOX 2.2: On Natural Capital[5]

Natural capital refers to any stock of natural assets that yields a flow of valuable goods and services into the future. For example, a forest, a fish stock or an aquifer can provide a harvest or flow that is potentially sustainable year after year. The forest or fish stock is "natural capital" and the *sustainable* harvest is "natural income." Natural capital also provides such services as waste assimilation, erosion and flood control, and protection from ultraviolet radiation. (Thus, the ozone layer is a form of natural capital.) These life-support services are also counted as natural income. Since the flow of services from ecosystems often requires that they function as intact systems, the structure and diversity of the system may be an important component of natural capital.

Researchers typically focus on three categories of natural capital: renewable, replenishable and non-renewable. Renewable natural capital, such as living species and ecosystems, is self-producing and self-maintaining using solar energy and photosynthesis. Replenishable natural capital includes surface and ground water supplies and the stratospheric ozone layer. These stocks are non-living but are continuously restored, often through some other solar mechanism. By contrast, non-renewable forms of natural capital such as fossil fuel and minerals are analogous to inventories. Any use implies liquidating part of the stock. Since adequate stocks of self-producing and replenishable natural capital are essential for life-support (and are generally non-substitutable), we consider these categories of natural capital to be more important to sustainability than non-renewable forms.

It should be apparent from the above that Earth's "natural capital" is more than just an inventory of industrial resources; it comprises also those components of the ecosphere, and the structural relationships among them, whose organizational integrity is essential for the continuous self-production and self-regulation of the system itself. Indeed, it is this highly evolved structural and functional integration that makes the ecosphere the uniquely livable "environment" it is. In effect the ecosphere is produced, in part, by the very organisms it comprises. In addition, geoclimatic, hydrological and ecological cycles do not simply transport and distribute nutrients and energy but are among the self-regulatory, homeostatic mechanisms that stabilize conditions on Earth for all contemporary life-forms, including humankind. All these, too, are forms of natural capital.

are scarier than anything the shift to sustainability might imply. Our increasingly global consumer lifestyle — living as if there were no biophysical limits to nature — not only undermines global life-support but also threatens geopolitical stability. In this context, the good news is that today so many people today accept the sustainability challenge as the first step toward a more secure future. The bad news is that the economic and political mainstream shows little sign of recognizing biophysical constraints of any kind. Indeed, "official" world development institutions seem more convinced than ever that the shortest route to sustainability is through unrestrained economic expansion.

In short, conflicting interests, opposing world views, incompatible analyses, rising material expectations, and fear of change, have led to a disorienting array of interpretations of sustainability and how to achieve it. Little wonder progress of any kind is slow! The problem is that not all interpretations of sustainability can be equally valid. The assumptions and the facts upon which each is based must be subject to logical scrutiny and repeated "reality checks" against empirical evidence before their prescriptions are accepted. In this light, let's examine more closely our own premise that humans must learn to live with each other within the means of nature.

Strong sustainability: the ecological bottom-line condition for sustainability

As long as Earth is humanity's only home, sustainability requires that we live within the productive capacity of nature. To use an economic metaphor, humankind must learn to live on the income generated by remaining natural capital stocks. "Natural capital" includes not only all the natural resources and waste sinks needed to support human economic activity, but also those biophysical processes and relationships among components of the ecosphere that provide essential life-support "services" (see Box 2.2).

If we consume more than the interest or income from our natural capital we diminish our biophysical wealth. This undermines our future because, despite our increasing technological sophistication, humans remain in a state of "obligate dependence" on the productivity and life-support services of the ecosphere.[6] Thus, from an ecological perspective, adequate land and associated productive natural capital are fundamental to the prospects for continued humane existence on Earth. Significantly, both the human population and average consumption are presently increasing while the total area of productive land and natural capital stocks are fixed or in decline.

These trends beg the question of how much natural capital is enough. Should we strive to conserve or enhance our natural capital stocks ("strong sustainability") or, as many economists believe, are losses of natural capital acceptable if compensated through the substitution of an equivalent amount or value of human-made capital ("weak sustainability" — see Box 2.3)?[7]

Certainly there are many examples of how technology has been able to substitute for natural resources. Microwave transmission and optical fibres have greatly reduced the demand for copper. However, we argue that in many situations the substitution option does not apply — natural capital (e.g., the forest) is often a prerequisite for manufactured capital (e.g., the sawmill). In other cases, technology and manufactured capital will simply not be able to substitute for critical natural capital (e.g., the ozone layer) in the foreseeable future. Even in the best of circumstances, therefore, blind faith in substitution

BOX 2.3: Strong or Weak Sustainability?[7]

Many economists believe that "weak sustainability" is good enough. According to this view, society is sustainable provided that the aggregate stock of manufactured and natural assets is not decreasing. In other words, weak sustainability allows the substitution of equivalent human-made capital for depleted natural capital. From this perspective, the loss of the income-earning potential of a former forest is no problem if part of the proceeds of liquidation have been invested in factories of equivalent income-earning potential. By contrast, "strong sustainability" recognizes the unaccounted ecological services and life-support functions performed by many forms of natural capital and the considerable risk associated with their irreversible loss. (In addition to wood fibre, forests provide flood and erosion control, heat distribution, climate regulation, and a variety of other non-market functions and values.) Strong sustainability therefore requires that natural capital stocks be held constant independently of human-made capital. Some authors suggest that manufactured capital stocks must also be held constant for strong sustainability so there is no capital depreciation of any kind. We agree that this is to be preferred, but wish to emphasize the greater importance of maintaining adequate life-supporting natural capital. Remember too that if population and material expectations are rising, capital stocks should actually be enhanced — in other words, it is *per capita* stocks that must be increased.

The weakness of "weak sustainability" is best revealed in a study by David Pearce and Giles Atkinson. Starting from the weak sustainability assumption that natural and human-made capital are substitutable, they ranked the sustainability of 18 representative countries. They propose that "...an economy is sustainable if it saves more [in monetary terms] than the depreciation on its [hu]man-made and natural capital...." As a result, Japan, the Netherlands and Costa Rica head the list of sustainable countries, while the poorest nations in Africa are identified as the most unsustainable. This comparison demonstrates the ecological irrelevance of "weak sustainability." It fails to recognize that much of the so-called rich countries' money savings comes from the depletion of other countries' natural capital and exploitation of global common-pool assets. For example, the apparent economic sustainability of both Japan and the Netherlands depends on large-scale imports (see Box 3.5). In effect, high material standards are maintained by a massive but unaccounted ecological deficit with the rest of the world (including some of the countries labeled "unsustainable").

would be a risky option. As things stand, the pace of stock depletion and accelerating global change suggests that remaining natural capital stocks are already inadequate to ensure long-term ecological stability. In these circumstances, we believe that "strong sustainability" is a necessary condition for ecologically sustainable development. More explicitly, this condition is met only if each generation inherits an adequate stock of essential biophysical assets that is no less than the stock of such assets inherited by the previous generation. (If today's average material standards are to be maintained, this "inheritance" will have to be on a *per capita* basis to keep ahead of population growth.) This version of the "constant capital stocks" condition is independent of the state of the human-made capital stock (although, if possible, the latter should also be held constant *per capita*).

However radical "strong sustainability" may appear as a conservation measure, the concept is still highly anthropocentric (human-centered) and narrowly functional. Emphasis is on the minimum biophysical requirements for *human survival* without regard to other species. Certainly too (as our more sensitive students like to remind us), we do not experience the taste, feel and smell — the sheer sensual exuberance — of nature as "natural capital." However, the preservation of biophysical assets essential to humankind does imply the direct protection of whole ecosystems and thousands of keystone species, and many other organisms would benefit indirectly as well. In short, the most promising hope for maintaining both significant biodiversity and the experience of nature under our prevailing value system may well be ecologically enlightened human self-interest. Of course, should humankind shift to more ecocentric values, its own survival might be assured even more effectively. Respect for, and the preservation of, other species and ecosystems for their intrinsic and spiritual values would automatically ensure human ecological security.

We must also recognize that maintaining the ecological bottom-line is not in itself sufficient for sustainability. Certain minimal socioeconomic conditions must also be met to ensure the necessary consensus for short-term action and long-term geopolitical stability. In the final analysis, sustainability means securing a satisfying quality of life for everyone. Most importantly, therefore, we must work to achieve basic standards of material equity and social justice both within and between countries (an objective which seems to be *receding* today). We also need a shared commitment to our collective interest in the maintenance of the global commons, an idea still struggling to be heard amidst the sterile rhetoric of competitive economic globalization. If we do not satisfy these conditions we simply will not be able to develop the capacity to tackle global change and the inevitable conflicts it generates in a humane and co-operative manner.

Having spelled out the bottom-line for sustainability, we need to focus on how to put these conditions into practice. But let us proceed cautiously into the fixing mode — after all, the causes of many of our present problems are yesterday's "quick-fix" solutions. Significantly, in this light, even the Brundtland Commission's suggestions favor the technological fix.

The Brundtland Commission's proposed response

Many analysts have argued that the "solutions" proposed by the WCED are inconsistent with its own definition of sustainability. In fact, the Commission was curiously ambiguous in elaborating on its definition.[8] *Our Common Future* defines "needs" as the "...essential needs of the world's poor, to which over-riding priority should be given...." It also acknowledges the "...limitations imposed by the state of technology and social organization on the environment's ability to meet [those needs]...." To people concerned about ecological integrity and social equity, this focus on "essential needs of the...poor" and "limitations" seemed to be a plea for political recognition of global economic injustice and limits to material growth. This guaranteed endorsement of *Our Common Future* by most mainstream environmental groups.

But there is another side to *Our Common Future* which guaranteed its message would be as enthusiastically received in corporate boardrooms everywhere. The report reassuringly asserts that "...sustainable development is not a fixed state of harmony, but rather a process of change in which the exploitation of resources, the orientation of the technological development, and institutional change are made consistent with future as well as present needs...." Indeed, close reading reveals that the only "limitations" recognized by the commission are social and technological. Achieving sustainable development is therefore said to depend on broader participation in decision-making; new forms of multilateral co-operation; the extension and sharing of new technologies; increased international investment; an expanded role for transnational corporations; the removal of "artificial" barriers to commerce; and expanded global trade.

In effect, the Brundtland Commission equated sustainable development with "...more rapid economic growth in both industrial and developing countries..." on grounds that "...economic growth and diversification...will help developing countries mitigate the strains on the rural environment...." Consistent with this interpretation, the commission observed that "...a five- to ten-fold increase in world industrial output can be anticipated by the time world population stabilizes some time in the next century...." While this may seem like an extraordinary rate of expansion, it implies an average annual growth rate in the vicinity of only 3.5 to 4.5 percent over the next 50 years. Growth in this range has already produced a five-fold increase in world economic output since the Second World War.

In recognition of the additional stress this expansion implies for the environment, the commission cast sustainable development in terms of more material- and energy-efficient resource use, new ecologically benign technologies and "...a production system that respects the obligation to preserve the ecological base for development...." Notably absent from *Our Common Future*, however, is any analysis of the causes of the poverty and inequity the Commission seeks to address or of whether the required growth would be biophysically sustainable under any conceivable production system. Nor did the commission confront arguments that under prevailing conditions liberalized trade and conventional efficiency gains may actually work against sustainability (see Chapter 4).

For such reasons, critics of the Brundtland Commission label its growth-dependent interpretation of sustainable development as a "...menace in as much as it has been co-opted [by the mainstream]...to perpetuate many of the worst aspects of the expansionist model under the masquerade of something new...." Even popular commentators condemn the use of the term "sustainable development" as "...dangerous words now being used...to mask the same old economic thinking that preaches unlimited consumption in the crusade to turn more land into glorified golf courses, deadly suburban ghettos, and leaking garbage holes (so-called landfill sites)....."[9]

Little wonder there is so much tension among various interests in their efforts to define sustainability and so much public disillusion with the concept. In today's materialistic, growth-bound world, the politically acceptable is ecologically disastrous while the ecologically necessary is politically impossible. Developing sustainability strategies that are consistent with the ecological bottom-line therefore depends on the convergence of ecological and political practicality. This is where the Ecological Footprint comes in: it is a consciousness-raising tool that can help us to develop a common understanding of the problem and explore the implications of alternative solutions. As such, it can help translate strong sustainability into planning action.

The Ecological Footprint: A Tool for Planning Toward Sustainability

Measuring progress toward sustainability: the dos and don'ts

Gaining acceptance for strong sustainability hinges on finding a meaningful unit to measure the natural capital requirements of the economy. Is nature's productivity sufficient to satisfy present and anticipated demands by the human economy indefinitely? This question seems so self-evidently crucial to sustainability that it is difficult to imagine how policy analysts in government, the private sector and universities can continue systematically to ignore it.

Part of the problem is that conventional economic models see the human economy as one in which the factors of production (e.g., labor, capital, information) are near perfect substitutes for one another and in which using any factor more intensely guarantees an increase in output. Any other resource limitation can be relieved by trade. In effect, this vision assumes a world with infinitely expandable carrying capacity.

Another difficulty is that conventional analysis is based on the circular flow

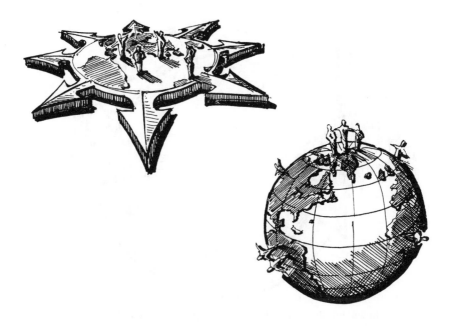

Figure 2.2: Flat-earth Versus Round-earth Economics.
Conventional economics is "flat-earth" economics. It implicitly sees the world extending without limit in all directions and imposing no serious constraints on economic growth. By contrast, ecological economics recognizes the world as a finite sphere. All resources come from the Earth and go back to it in degraded form. The only "income" from outside is sunlight, which powers material cycles and the web of life. Economic activity is therefore ultimately constrained by the regenerative capacity of the ecosphere.

of exchange value (money flows) between households and firms and back again as exemplified by standard measures of GDP. Physical measures of natural capital, natural income, and subsequent energy/material transformations are simply not part of the analysis (Figure 2.3). Indeed, mainstream models of growth and sustainability lack any representation of the biophysical "infrastructure" and the time-dependent processes upon which the economy depends and which are basic to an ecologically informed approach (Figure 2.4). Most important, there is no reference to modern interpretations of the Second Law of Thermodynamics which see the economy as a complex "dissipative structure" embedded within the ecosphere (see Box 2.4 for a more detailed explanation). Many critical questions raised by ecological and thermodynamic considerations are therefore invisible to mainstream approaches. It seems that

Figure 2.3: The economic perspective: circular flows. Conventional economics emphasizes the seemingly self-generating circular flows of money between firms and households in the marketplace. It thereby fails to account for either informal work or the value of ecological services, and is blind to the irreversible unidirectional material flows that sustain the economy.

conventional indifference to carrying capacity derives not from superior knowledge but, from conceptual weaknesses in standard analytic models.

One can monitor the availability of energy, matter, and other forms of natural income either in terms of *physical* measurements of natural capital stocks and flows or in terms of *monetary* measurements such as the dollar value of stocks and current prices for marketed goods and services. No doubt, money prices are critical for operating in the public domain. Financial analysis is crucial when developing budgets, or when deciding between building a school, a hospital or a theatre; business decisions are unthinkable without sound monetary analyses. However, we argue that monetary analyses are fatally flawed in assessing sustainability issues or natural capital constraints. Using money price to signal resource scarcity or natural capital depletion may be misleading for at least the following reasons (Figure 2.5)[10]:

One: Monetary interpretations of the constant natural capital requirement may mask declining physical stocks. For example, some economists suggest that the constant capital stocks condition for sustainability might be satisfied if the money value of, or income from, capital is held more or less constant. According to neoclassical theory, the marginal price of increasingly scarce resource commodities should increase. If this premise holds true, rising prices (which should indicate increased scarcity) could hold the income from, or the

---

Box 2.4: The Entropy Law
and the Economy/Ecology Conundrum

The Second Law of Thermodynamics (the "entropy law") states that the entropy of an isolated system always increases. This means that the system spontaneously runs down. All available energy is used up, all concentrations of matter are evenly dissipated, all gradients disappear. Eventually, there is no potential for further useful work — the system becomes totally degraded and "disordered." This has significant implications for sustainability:

» Non-isolated systems (such as the human body or the economy) are subject to the same forces of entropic decay as are isolated ones. This means that they must constantly import high-grade energy and material from the outside, and export degraded energy and matter to the outside, to maintain their internal order and integrity. For all practical purposes, this energy and material "throughput" is unidirectional and irreversible.

» Modern formulations of the Second Law therefore argue that all highly-ordered, far-from-equilibrium, complex systems necessarily develop and grow (increase their internal order) "at the expense of increasing disorder at higher levels in the systems hierarchy."*

» The human economy is one such highly-ordered, complex, dynamic system. It is also an open sub-system of a materially closed, non-growing ecosphere, i.e., the economy is *contained* by the ecosphere. Thus the economy is dependent for its maintenance, growth and development on the production of low entropy energy/matter (essergy) by the ecosphere and on the waste assimilation capacity of the ecosphere.

» This means that beyond a certain point, the continuous growth of the economy (i.e., the increase in human populations and the accumulation of manufactured capital) can be purchased only at the expense of increasing disorder (entropy) in the ecosphere.

» This occurs when consumption by the economy exceeds production in nature and is manifested through the accelerating depletion of natural capital, reduced biodiversity, air/water/land pollution, atmospheric change, etc.

* E. Schneider and J. Kay. 1992. *Life as a Manifestation of the Second Law of Thermodynamics.* Preprint from: *Advances in Mathematics and Computers in Medicine.* (Waterloo, Ont.: University of Waterloo Faculty of Environmental Studies, Working Paper Series).

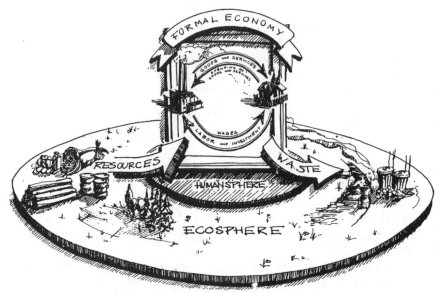

Figure 2.4: The ecological worldview.
The circular flows are actually sustained by the unidirectional throughput of
ecological goods and services from and to the ecosphere (the "natural income"
stream). All the energy and much of the matter that passes through the economy
is permanently dissipated into "the environment" never to be used again.

total value of, a particular natural capital stock constant, while the physical
stock is actually in decline. Thus, constant money income or stock value may
foster the illusion of constant stocks while physical inventories shrink. In other
cases, prices may fall (suggesting resource abundance) while the stock is
depleted due to extra-market factors or improved extraction technology (as
illustrated by mineral and fossil fuel prices in recent decades). In either case,
market prices would mask the depletion of stocks.

Two: In any event, biophysical or eco-functional scarcity is poorly reflected
in the marketplace. Market prices generally say nothing about the size of
remaining natural capital stocks or whether there is some critical minimal stock
size below which recovery is impossible. In short, prices do not monitor stock
size or systems fragility, but only the commodity's short-term scarcity on the
market. Even this is not quite true — market prices are more influenced by
short-term demand; the state of technology (extraction, processing and trans-
action costs); the intensity of competition; the availability of substitutes; etc.,
than by market scarcity. For example, subsidies, low fuel costs, and high-tech
factory freezer-trawlers enabled industrial fishers to access previously un-
reachable stocks of North Atlantic groundfish. This maintained market supply
(and relatively low prices) even as the stocks were being depleted. In any case,

Figure 2.5: Measuring the World in Monetary Units Makes us Blind to the Ecological Constraints on Sustainability. Acknowledging the limitations of monetary assessments becomes an additional argument against "weak sustainability." As noted earlier, the weak criterion assumes the substitut-ability of human-made for natural capital, allowing (false) "trade-offs" in terms of equivalent stock values or income-generating potential. An alternative approach is to assess our natural capital requirements from an ecological and biophysical perspective.

fish prices have to compete with those of pork and chicken (substitutes) and do not sky-rocket even in the event of a fishery collapse.

Any remaining value of price as an indicator of scarcity of biophysical stocks is undermined by the behavior of complex systems. Conventional models assume smooth, reversible change in supply and prices. In fact, natural systems are more likely to be characterized by time lags and sudden irreversible change (or very long recovery times), systems behaviors that cannot be detected in the market.

Three: Monetary analyses are systematically biased against the future by discounting. Consider that at a discount rate of 5 percent, the present value of a dollar's worth of ecological service a life-span (76 years) from now is only about 2.5 cents. In other words, 2.5 cents put in the bank today at 5% would grow to about one dollar in 76 years. Discounting makes nature appear less valuable the further into the future we look. However, life depends on ecologi-cal continuity: for all we know, future generations will need the same amount of the same kinds of critical ecological goods and services *per capita* as we do today, whatever the discounted present (money) value of those goods. Never-theless, we regularly sacrifice nature to development because the immediate short-term benefits exceed the (discounted) present value of foregone future benefits — or at least our estimate of what they will be. For example, paving over agricultural land for a shopping center today assumes that we know both the future value of the lost ecological productivity and that anticipated money profits will more than compensate for this loss. Both assumptions are increas-ingly risky in today's uncertain world. The value of human-made capital (the shopping center) today tells us less and less about its potential money income and nothing about the demand for food (natural income) tomorrow. The value

Figure 2.6: Carrying Capacity is traditionally defined as the maximum population of a species that can be sustained indefinitely in a given habitat.

of natural capital to human life will almost certainly increase more rapidly than that of manufactured capital over time as evidence of ecological breakdown becomes more compelling, whatever today's markets tell us. (For example, the effective price of the stratospheric ozone layer went from zero to near infinity in just a few years in the absence of any market.) In this light, standard approaches to discounting nature's services constitute a dangerous systematic bias against the future.

Four: The utility of monetary indicators is further diminished by market fluctuations, which affect prices but not the ecological value or integrity of natural capital. For example, world price fluctuations are unrelated to local circumstances or inter-regional variations, yet affect the relative economic strength of different regions and with it perceived values of local natural capital. Money values and markets may therefore seriously alter local conservation and management practices respecting agricultural land, for example, even though its inherent productivity and potential contribution to long-term food security remains unchanged.

Five: Money values do not distinguish between substitutable goods and complementary goods. Moreover, on monetary balance sheets, all prices are added or subtracted as if goods that are priced the same are of equal importance to human life —money equivalency equates the essential with the trivial. In fact, of course, many goods and services of nature are virtual prerequisites to life and therefore are not truly commensurate with some human-made gadget of equal dollar value. While there certainly is substitutability between various industrial resource inputs (glass fibres are replacing copper cables in communications and data transmission), this functional equivalency does not apply to all potential natural and manufactured capital trade-offs. In some cases, once nature has been over-exploited, no amount of manufactured goods will compensate for the loss of natural capital. To put fish on our dinner plates, both a fish stock *and* fishing boats are needed. Thus, even though the fishing fleet and canning factories may be worth as much in dollars as the fish stock, all the fishing equipment and processing plants in the world will not generate a single fish if the natural stock is destroyed. In short, more often than not, natural capital is a prerequisite for human-made goods, while the opposite is not the case.

Six: The potential for growth of money is theoretically unlimited, which obscures the possibility that there may be biophysical limits to economic growth. To use Herman Daly's metaphor, monetary analysis does not recognize the Plimsoll line, which indicates the maximum load capacity of a ship. Overloading (excessive growth) may eventually sink the ship. Pareto efficiency — the current criterion of macro-economic health — ensures only that the load is distributed in such a way that the ship sinks optimally!

Seven: Perhaps the most serious objection is that there are no markets for many critical natural capital stocks and life-support processes (e.g., the ozone layer, nitrogen fixation, global heat distribution, climatic stability, etc.). Conventional approaches to conservation and sustainability focus mainly on the money values of marketable resource commodities (e.g., timber and wood fibre) and are insensitive to the intangible (but ultimately more valuable) non-market functions of the natural capital that produces them (e.g., the forest ecosystem). The latter functions are destroyed by resource harvesting. Not surprisingly, therefore, economists today are devoting much attention to ways of "putting a price on nature." However, there are severe limitations on the possibilities of establishing valid shadow prices even for familiar ecological goods and services, and no possibility at all for those many functions whose existence is unknown (and may be inherently unknowable) before some breakdown occurs. In these circumstances, prices fail absolutely as scarcity indicators.

In summary, monetary approaches are blind to the requirements for ecological sustainability because they do not adequately reflect biophysical scarcity, social equity, ecological continuity, incommensurability, structural and functional integrity, temporal discontinuity, and complex systems behavior.

Learning from ecology: revisiting human carrying capacity

The renewed debate around the natural capital constraints on the economy demands that we revisit the ecological concept of carrying capacity.[11] Does it make sense to talk of the *human* carrying capacity of Earth? For purposes of game and range management, carrying capacity is usually defined as the

---

BOX 2.5: A Brief History of the Human Carrying Capacity Concept [13]

The oral history of concern about the relationship between people and land must go back thousands of years. Many Chinese and early Christian scholars worried about the destruction of habitat. Plato may have provided the first written account of human carrying capacity as he declared in *Laws*, Book V, that a

> ...suitable total for the number of citizens cannot be fixed without considering the land and the neighboring states. The land must be extensive enough to support a given number of people in modest comfort, and not a foot more is needed.

The first scholarly book on sustainable practice in the English language may be John Evelyn's *Sylva: A Discourse of Forest, Trees and the Propagation of Timber* published in 1664, two hundred years before George Perkins Marsh's study *Man and Nature* initiated the scientific debate in North America on nature's limited capacity to satisfy human demands.

Ecological accounting, the basis for carrying capacity assessments, can be traced back to at least as early as 1758. In that year, François Quesnay published his *Tableau Economique* in which the relationship between the productivity of land and wealth creation was discussed. Since then, many scholars have developed conceptual approaches and accounting procedures to analyze the relationship between people and nature.

Some of them looked at energy flows needed to support human activities. For example, in 1865 economist Stanley Jevons in *The Coal Question*, analyzed the importance of energy resources for the United Kingdom's economic performance. In the late 1800s, Serhii Podolinsky initiated the field of agricultural energetics. In the following decades, the eminent physicists Rudolf Clausius and Ludwig Boltzmann, and later Nobel Laureate Frederick Soddy, reflected upon the implications of the entropy law on economic development. Alfred Lotka introduced energy analysis to biology in the 1920s, and in the 1970s economist Nicholas Georgescu-Roegen challenged economics using thermodynamic principles.

Others have more explicitly examined the carrying capacity requirements of economies. For example, with his 1798 *Essay on the Principles of Population as It Affects the*

maximum population of a given species that can be supported indefinitely in a specified habitat without permanently impairing the productivity of that habitat. However, because of our seeming ability to increase human carrying capacity by eliminating competing species, by importing locally scarce resources, and through technology, this definition does not seems applicable to humans. Indeed, trade and technology are often cited as reasons for rejecting

---

Future Improvement of Society, Reverend Thomas Malthus initiated the debate on agriculture's seemingly limited ability to feed an ever larger human population. Even the Ecological Footprint has conceptual predecessors; in his above-mentioned book, Stanley Jevons observed that:

> the plains of North America and Russia are our [British] corn-fields; Chicago and Odessa our granaries; Canada and the Baltic are our timber-forests; Australasia contains our sheep-farms, and in Argentina and on the western prairies of North America are our herds of oxen; Peru sends her silver, and the gold of South Africa and Australia flows to London; the Hindus and the Chinese grow tea for us, and our coffee, sugar and spice plantations are all in the Indies. Spain and France are our vineyards and the Mediterranean our fruit garden, and our cotton grounds, which for long have occupied the Southern United States, are now being extended everywhere in the warm regions of the Earth.

Forty years later, in 1902, physicist Leopold Pfaundler calculated global carrying capacity, concluding that as an upper limit, ecological production could sustain about five people per hectare of land. In North America, William Vogt (1948) and Fairfield Osborn (1953) are associated with the renewed academic interest in carrying capacity questions. Georg Borgstrom in his various publications in the 1960s and early 1970s analyzed resource consumption in terms of "ghost acreage," which referred to imported agricultural carrying capacity. One of us (Rees) developed the "regional capsule" (subsequently the Ecological Footprint) concept in the early 1970s as a teaching tool to stimulate multi-disciplinary planning students to think about human carrying capacity. In 1980, William Catton added a new dimension to the human carrying capacity debate by describing the implications of overshoot — temporarily exceeding the long-term carrying capacity — and the subsequent population crash. G. Higgins and his collaborators produced a technical report in 1983 analyzing the population-supporting capacities of most developing countries for the United Nations Food and Agriculture Organization (FAO). In 1985, Ragnar Overby, then at the World Bank, proposed comparing economies by their demand on carrying capacity, and in 1986 M.A. Harwell and T.C. Hutchinson analyzed the loss of carrying capacity that would follow nuclear war. Most recently (1993) the Friends of the Earth (Netherlands) proposed the "environmental space" concept to help determine nations' fair shares of global productive/assimilative capacity.

These are only a few examples from the literature on human carrying capacity.

the concept of human carrying capacity out of hand.

This is an ironic error — shrinking carrying capacity may soon become the single most important issue confronting humanity. The reason for this becomes clearer if we define carrying capacity not as a maximum population but rather, following William Catton, as the maximum "load" that can safely and persistently be imposed on the ecosphere by people. Human load is a function not only of population but also of *per capita* consumption and the latter is increasing even more rapidly than the former due (ironically) to expanding trade and technology. This led Catton to observe that "...the world is being required to accommodate not just more people, but effectively 'larger' people...."[12] As a result, load pressure relative to carrying capacity is rising much faster than is implied by mere population increases.

These trends underscore the fact that despite our technological, economic and cultural accomplishments, human beings remain ecological beings. Like all other species we depend for both basic needs and the production of artifacts on energy and material resources extracted from nature. All this energy and matter is eventually returned to the ecosphere as waste. A full understanding of the human ecological "niche" must therefore include full consideration of the flows of available energy and matter into the economy and the return flows of degraded energy and material (wastes) back to the ecosystem.

Analysis of this biophysical "throughput" shows that humankind, through the industrial economy, has become the dominant consumer in most of the Earth's major ecosystems. By 1986, humankind — one species among millions — was already "appropriating," directly and indirectly, 40 percent of the net product of terrestrial photosynthesis and recent studies suggest that the human "take" from rich coastal marine environments is approaching 30 percent[14] (which may be beyond the sustainable yield — despite increasing effort, the world's fisheries catch has declined since 1989). What are the implications of such dominance for ecosystems integrity? Can it be safely extended? (Remember the North Atlantic groundfish stocks!) Meanwhile, such trends as ozone depletion and greenhouse gas accumulation show that critical global waste sinks are also filled to overflowing. All such data indicate that even today's levels of appropriation are unsustainable. The human "load" has grown to the point where total consumption already exceeds sustainable natural income.

Achieving ecological sustainability clearly requires that economic assessments of the human condition be based on, or at least informed by, ecological and biophysical analyses. The fundamental ecological question for ecological economics is whether remaining species populations, ecosystems and related biophysical processes (i.e., critical self-producing "natural capital" stocks), and the waste assimilation capacity of the ecosphere are adequate to sustain the anticipated load of the human economy into the next century while simultaneously maintaining the general life-support functions of the ecosphere. This critical question is at the heart of ecological carrying capacity but is virtually ignored by mainstream approaches.[15]

Turning carrying capacity on its head: human Ecological Footprints

Determining the human population that a given region might support is problematic for two major reasons: first, the total ecological load imposed by any population will vary with such factors as average income, material expectations, and the level of technology (e.g., energy and material efficiency). In short, human carrying capacity is as much a product of cultural factors as it is of ecological productivity. Second, in a global economy, no region exists in isolation — people have access to resources from all over the world. Indeed, as previously noted, many people argue that trade overcomes any regional limits to growth imposed by local resource shortages.

Other factors further complicate the carrying capacity question. Unlike consumption by other animals, consumption by people is not determined solely by biology. Because of technology, the load imposed by our biological metabolism is vastly augmented by industrial metabolism. While most species consume little beyond their food, the bulk of human material consumption consists of manufactured non-food items such as energy, clothing, automobiles and a vast array of other consumer goods. In industrialized countries, such material consumption is positively encouraged by the culture of consumerism, and constrained only by spending power. Globally, of course, individual consumption levels vary by orders of magnitude: farm-hands in rural India might represent the lower extreme of the scale, board members of transnational companies the upper echelon.

Ecological Footprint analysis gets around some of the difficulties with "traditional" carrying capacity simply by inverting the usual carrying capacity ratio. The Ecological Footprint starts from the assumption that every category of energy and material consumption and waste discharge requires the productive or absorptive capacity of a finite area of land or water. If we sum the land requirements for all categories of consumption and waste discharge by a defined population, the total area represents the Ecological Footprint of that population on the Earth *whether or not this area coincides with the population's home region*. In short, the Ecological Footprint measures land area required per person (or population), rather than population per unit area. As we shall see, this simple inversion is far more instructive than traditional carrying capacity in characterizing the sustainability dilemma.

More formally, the Ecological Footprint of a specified population or economy can be defined as the area of ecologically productive land (and water) in various classes — cropland, pasture, forests, etc. — that would be required on a continuous basis

a) to provide all the energy/material resources consumed, and

b) to absorb all the wastes discharged
by that population with prevailing technology, *wherever on Earth that land is located*. Consumption by households, businesses and governments is included in the calculations. Note that because the Ecological Footprint is based on natural income flows, it also provides an area-based estimate of the natural capital requirements of the subject population.

As suggested above, the size of the Ecological Footprint is not fixed but is dependent on money income, prevailing values, other sociocultural factors, and the state of technology. Keep in mind, however, that whatever the specifics, the Ecological Footprint of a given population is the land area needed *exclusively* by that population. Flows and capacities used by one population are not available for use by others.

Complete Ecological Footprint analysis would include both the direct land requirements and indirect effects of all forms of material and energy consumption. Thus, it would include not only the area of different ecosystems (natural capital) required to produce renewable resources and life-support services (different forms of natural income) but also the land area lost to biological productivity because of contamination, radiation, erosion, salination, and urban "hardening" — the paving over or building up of land that makes it ecologically unproductive. It would also factor in non-renewable resource use insofar as it can account for processing energy and use-related pollution effects. At present, however, our assessments are based on a limited range of consumption items and waste flows. Every additional item would therefore increase the size of our existing estimates. In addition, the present calculations assume that the required land (e.g., in forestry or agriculture) is being used sustainably. However, this is not generally the case — croplands, for example, are typically degraded 10 times faster than they can regenerate. This means that although the calculated Ecological Footprints for industrial regions and countries are impressively large, they are, if anything, considerable under-estimates of the effective demand. The case could be made that our present estimates should be increased by a significant "sustainability factor" to account for such simplifying assumptions.

"Turning carrying capacity on its head" eliminates several objections to the application of the concept to humans. It is true, as critics claimed, that trying to measure human carrying capacity in terms of maximum supportable regional population is a futile exercise. Local populations are so influenced by culture, trade and technological factors that any relationship to local biophysical limits is obscured. Hong Kong, for example, is densely populated and wildly prosperous yet has very little natural carrying capacity, while many African countries with much larger biophysical capacities suffer from famine. The Ecological Footprint gets around this analytic problem by measuring the population's total load rather than the number of people. This recognizes that

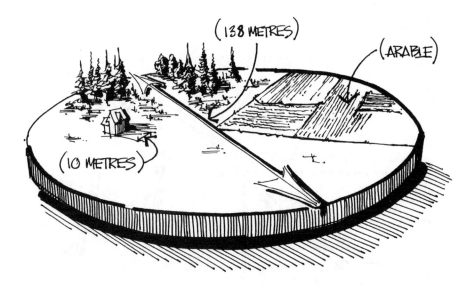

Figure 2.7: A fair Earthshare is the amount of land each person would get if all the ecologically productive land on Earth were divided evenly among the present world population. If your present Earthshare were a circular island it would have a diameter of just 138 metres. One sixth of your island would be arable land, the rest pasture, forest and wilderness, and built-up area. Clearly, as the population increases, our earthshares shrink. Also, for each person whose Ecological Footprint exceeds his/her fair earthshare by, say, a factor of three (as do North Americans'), three other people would have to content themselves with only a third of a share for global sustainability. —Any volunteers?

people have an impact somewhere even if it is obscured by trade and technology. Indeed, to the extent that trade seems to increase local carrying capacity, *it reduces it somewhere else.*

Our method summarizes a given population's impacts on nature by analyzing aggregate consumption (i.e., total load = population x *per capita* consumption) and converting this to a corresponding land area. We can thus produce a single measure of ecological demand (or natural capital requirements) which, unlike traditional carrying capacity, accounts for net trade and reflects both current income and prevailing technology. The Ecological Footprint so calculated can be compare to the area of the population's home region to reveal the extent to which local carrying capacity has been exceeded and therefore the population's dependence on trade. (Bits of a population's Ecological Footprint can be all over the world.) The Ecological Footprint also facilitates comparison between regions and thus reveals the effect of differing income levels and technology on ecological impact. It should be no surprise that while local capacity is severely limited, the Ecological Footprint of the

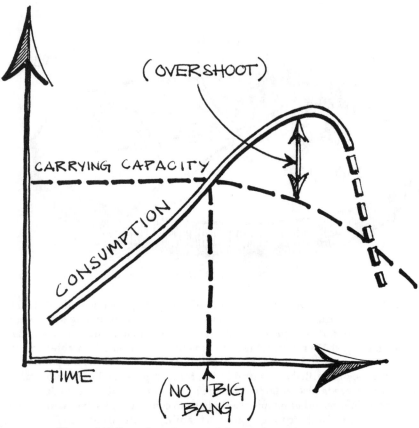

Figure 2.8: Overshoot is growth beyond carrying capacity.
Carrying capacity limits can be overshot without a "big bang" because of the availability of large capital stocks. Harvests can still increase and money incomes rise, and while there may be indications of ecological stress, all else may seem normal. Ultimately, however, the consequences of eroded natural capital may be felt as eco-catastrophe and population crash.

average Hong Kong resident is vastly larger than that of an Ethiopian farmer!

The Ecological Footprint approach can be adapted to other sustainability assessments. For example, we could compute the Ecological Footprint of trade to reveal how much "carrying capacity" is embodied in a region's imports and how much capacity it gives up to produce the exports required to pay for its imports. Also, individual or average *per capita* Ecological Footprints[16] can be compared to the current *fair Earthshare*. A fair Earthshare is the amount of ecologically productive land available per person on Earth (with apologies to other species!). Today, this amounts to 1.5 hectares (3.7 acres), or a 122 metre square. Only 0.25 hectares (0.62 acres) of this are arable (Figure 2.7).

Perhaps importantly, Ecological Footprint analysis allows us to estimate the extent of global *overshoot* and the *ecological deficit* of any specified region or country. "Overshoot" is the amount by which humanity's total Ecological Footprint is bigger than global carrying capacity (see Figure 2.8). Beyond a certain point, the material growth of the world economy can be purchased only at the expense of depleting natural capital and undermining the life-support services upon which we all depend. In other words we are in overshoot when consumption by the economy exceeds natural income as indicated by ecological decline. The ecological or sustainability deficit is a measure of "local" overshoot. It estimates the difference between a defined region's or country's domestic ecological capacity and its actual Footprint. It therefore reveals the extent to which that region is dependent on extra-territorial productive capacity through trade or appropriated natural flows.

There is much evidence today that humanity's Ecological Footprint already exceeds global carrying capacity. Such overshoot is only possible temporarily and imposes high costs on future generations. Without a concerted effort today to reduce material throughput, our children will have to satisfy the natural income requirements and other needs of a larger population from much-diminished stocks of natural capital (real wealth).

How Ecological Footprint analyses can help advance sustainability

Measuring a wide array of human activities in terms of their Ecological Footprints makes it possible to compare their separate ecological impacts. At the same time, in contrast to conventional "one-shot" environmental assessments, Footprint analysis allows a cumulative approach to impact analysis. Every economic activity imposes a demand on the ecosphere and the Ecological Footprint shows how all these demands for food and fibre, non-renewable resources, waste absorption, urban development, and even maintaining biodiversity, compete for ecological space. (The expansion of the human enterprise necessarily "appropriates" resources and habitat from other species.)

The Earth is wondrously productive and has an enormous capacity to support people and their economies, to say nothing of other species. However, the production of many goods and services in the increasingly global marketplace is already based as much on natural capital depletion, including important self-producing forms such as fish stocks, as it is on sustainable flows. The Ecological Footprint concept is an effective tool in raising this emergent reality to consciousness. It is unfortunate that neither price nor product labels declare whether our consumption goods represent interest or the drawing down of nature's savings.

Using productive land area as a measurement unit makes Ecological Footprint analysis consistent with basic laws of physics, especially the laws of mass balance and thermodynamics. In particular, the modern world has to come to grips with the second-law axiom that any complex, self-organizing system (such as the economy) must have a continuous input of energy and matter from its "host" system to grow and maintain itself (i.e. to overcome its internal entropic decay — see Box 2.4). In this light, land or ecosystem area is a more appropriate accounting unit for the human economy than energy flux alone because it reflects both the quantity and quality of energy and matter available to the human economy. The key limiting factor for human life is not the amount of solar energy that falls on Earth, but what nature can do with it. For example, one lonely plant growing on one hectare of the Sahara desert is ecologically and economically less significant than one hectare of tropical forest, even though both receive the same solar radiation.

This last point emphasizes that the attributes of "land" go beyond the laws of thermodynamics. Land area not only captures planet Earth's finiteness, it can also be seen as a proxy for numerous essential life-support functions from gas exchange to nutrient recycling. The state of the biophysical world can therefore best be estimated from the state of the self-producing natural capital stocks that perform these functions. Keep in mind that these stocks themselves represent the biochemical energy that has accumulated in the ecosphere. The point is that land supports photosynthesis, the energy conduit for the web of life. This singular process distinguishes our planet from dead ones like Mars or Venus. Photosynthesis sustains all important food chains and maintains the structural integrity of ecosystems. It has miraculously transformed the originally inhospitable surface of the Earth into a self-producing and self-regulating ecosphere of spectacular abundance and diversity.

The Ecological Footprint reminds us that regardless of technology, humankind remains dependent on ecological goods and services and that these must be available in *increasing* quantities from somewhere on the planet as human populations and *per capita* consumption grow. As noted previously, the fundamental ecological question for sustainability is whether stocks of natural capital will be adequate to meet anticipated demand. Ecological Footprint analysis approaches this question directly. It provides a means to compare production by the ecosphere with consumption by the economy, thereby revealing whether there is ecological room for economic expansion or, on the other hand, whether industrialized societies have overshot local (and global) carrying capacity. In the latter case, the Ecological Footprint also reveals the sustainability gap confronting society. In short, Ecological Footprint analysis can help to determine the ecological constraints within which society operates; to shape policy to avoid or reduce overshoot; and to monitor progress towards achieving sustainability.

Ecological Footprint analysis by no means implies that living at carrying capacity is a desirable target. Rather, the Ecological Footprint is intended to show how dangerously close we have come to nature's limit. Ecological resilience and social well-being are more likely to be assured if the total human load remains substantially *below* Earth's carrying capacity. Living at the ecological edge compromises ecosystems' adaptability, robustness, and regenerative capacity, thereby threatening other species, whole ecosystems, and ultimately humanity itself.

Recognition of biophysical limits and the fact that human uses of nature are competing raises pertinent social and economic questions. For example, it forces over-consumers to face the otherwise hidden relationships and implicit trade-offs between their wealth and the poverty and human suffering that persists elsewhere. If these biophysical limits are real, should not mechanisms for redistribution be as prominent as economic efficiency and expansion are in plans to combat growing material inequity? Recognition that not everyone can become as materially rich as today's North Americans or Europeans without undermining global life-support should impose greater accountability on the wealthy and give the poor greater leverage in bargaining for development rights, technology transfers and other equity measures. Ecological Footprint analysis might therefore strengthen the case for international agreements on how to share the global commons and the Earth's productive capacity more equitably and how to use it more carefully.

The discussion so far has been relentlessly anthropocentric. However, Ecological Footprints also raises to consciousness humanity's disproportionate appropriation of energy/material flows and habitat that otherwise would be available for other species. Do we have an inherent right to so much of nature's productivity at the expense of the several million other species living on the planet?

In summary, by putting sustainability in simple but concrete terms, the Ecological Footprint concept provides an intuitive framework for understanding the ecological bottom-line of sustainability. This in turn stimulates public debate, builds common understanding and suggests a framework for action. The Ecological Footprint makes the sustainability challenge more transparent — decision-makers have a physical criterion for ranking policy, project, or technological options according to their ecological impacts. Finally, the Ecological Footprint underscores the global imperative for local action. It demonstrates that the ecological and social impacts of over-consumption reach far beyond our home regions. This introduces the moral dimension of sustainability and, by showing the contribution of both population growth and material consumption to global decline, emphasizes the need for policies to address them both. The following chapter describes specific applications of Ecological Footprint analysis.

Notes

1. For a more detailed discussion of these trends consult the Worldwatch Institute's annual *State of the World* and *Vital Signs* (NY: W.W. Norton), or the World Resources Institute, UNEP and UNDP's biannual *World Resources* (NY: Oxford University Press).
2. Worldwatch Institute, *Vital Signs* (NY: W.W. Norton, 1995)
3. The World Commission on Environment and Development (WCED) was chaired by Norwegian Prime Minster Gro Harlem Brundtland. The opening statement is from page 27 and the sustainable development definition from page 43 of its report, *Our Common Future* (NY: Oxford University Press, 1987).
4. Sustainability definitions are discussed in Sharachchandra M. Lélé, "Sustainable Development: A Critical Review," *World Development* Vol.19, No.6 (1991): 607-621; in the annex of David Pearce, Anil Markandya and Edward Barbier, *Blueprint for a Green Economy* (London: Earthscan Publications, 1989); and in William E. Rees, *Defining Sustainable Development* (The University of British Columbia, Vancouver: Centre for Human Settlements Publication, 1989). Further see Herman E. Daly, "Elements of Environmental Macroeconomics," in Robert Costanza, ed., *Ecological Economics: The Science and Management of Sustainability* (NY: Columbia University Press, 1991); Lester W. Milbrath, *Envisioning a Sustainable Society: Learning Our Way Out* (Albany, NY: State University of New York Press, 1989); and Michael Redclift, *Sustainable Development: Exploring its Contradictions* (London: Methuen & Co., 1987).
5. Liberally adapted from Robert Costanza and Herman E. Daly, "Natural capital and sustainable development," *Conservation Biology* Vol.1 (1992): 37-45 and William E. Rees, "Achieving Sustainability: Reform or Transformation?," *Journal of Planning Literature* Vol.9, No.4 (1995).
6. William E. Rees makes the case for "obligate dependence" in "Sustainable Development and the Biosphere: Concepts and Principles," *Teilhard Studies Number 23* (Chambersburg, PA: Anima Books (for American Teilhard Association for the Future of Man), 1990).
7. The weak-strong distinction was brought forward by both David Pearce *et al.* (1989, see above) and Herman Daly and John Cobb (*For the Common Good*, Boston: Beacon Press, 1989). Documents refuting the sustainability crisis are Marcus Gee, "Apocalypse Deferred: The End Isn't Nigh," in *The Globe and Mail*, 9 April 1994, D1-3; and, Julian L. Simon and Herman Kahn, eds., *The Resourceful Earth: A Response to Global 2000* (NY: B. Blackwell, 1984). David Pearce and Giles Atkinson's study is called "Capital Theory and the Measurement of Sustainable Development: An Indicator of 'Weak' Sustainability" in *Ecological Economics* Vol.8, No.2 (1993): 103-108.
8. The following quotes are from *Our Common Future*, pages 43, 9, 89, 213 and 65.
9. Duncan M. Taylor, "Disagreeing on the Basics: Environmental Debates Reflect Competing Worldviews," in *Alternatives* Vol.18, No.3 (1992): 26-33; and A. Nikiforuk, "Deconstructing Ecobabble: Notes on an Attempted Corporate Takeover," *This Magazine* Vol.24, No.3 (1990): 12-18.
10. References and complementary readings include: Herman E. Daly and Kenneth N. Townsend, eds., *Valuing the Earth: Economics, Ecology, Ethics* (Cambridge, MA: The MIT Press, 1993); Charles A.S. Hall, "Economic Development or Developing Eco-

nomics: What Are Our Priorities," in Mohan K. Wali, *Ecosystem Rehabilitation, Volume 1: Policy Issues* (The Hague, the Netherlands: SPB Academic Publishing, 1992); Colin Price, *Time, Discounting and Value* (Oxford: Blackwell Publishers, 1993); Andrew Stirling, "Environmental Valuation: How Much is the Emperor Wearing?," *The Ecologist* Vol.23, No.3 (1993): 97-103; and Arild Vatn and Daniel W. Bromley, "Choices without Prices without Apologies," *Journal of Environmental Economics and Management* 26 (1994):129-148.

11. Abstracted in part from William E. Rees, "Revisiting Carrying Capacity: Area-Based Indicators of Sustainability," *Population & Environment* (1995, in press).

12. William Catton, "Carrying Capacity and the Limits to Freedom," paper prepared for the Social Ecology Session, *XI World Congress of Sociology*, New Delhi, India, 18 August 1986.

13. Plato in David F. Durham, "Carrying Capacity Philosophy," *Focus* Vol.4, No.1 (1994): 5-7; early Christian and Chinese scholars in William Ophuls and A. Stephen Boyan Jr., *Ecology and the Politics of Scarcity Revisited*, (NY: W.H. Freeman and Company, 1992) (original edition 1977); John Evelyn in James Garbarino, *Toward a Sustainable Society: An Economic, Social, and Environmental Agenda for our Children's Future* (Chicago: The Noble Press Inc., 1992); Alfred James Lotka, *Elements of Physical Biology* (Baltimore: Williams & Wilkins, 1925); Nicholas Georgescu-Roegen, *The Entropy Law and the Economic Process* (Cambridge, MA: Harvard University Press, 1971); Leopold Pfaundler, "Die Weltwirtschaft im Lichte der Physik" [The Global Economy from the Point of View of Physics], in *Deutsche Revue*, Richard Fleischer, ed., Vol.27, No.2 (April-June 1902): 29-38,171-182; William Vogt, *Road to Survival* (NY: William Sloane, 1948); Fairfield Osborn, *The Limits of the Earth* (Boston: Little, Brown and Co., 1953); Georg Borgstrom, *Harvesting the Earth* (NY: Abelard-Schuman, 1973); William E. Rees, "An Ecological Framework for Regional and Resource Planning" (The University of British Columbia, Vancouver: UBC School of Community and Regional Planning, 1977); William R. Catton, *Overshoot: The Ecological Basis of Revolutionary Change* (Urbana: University of Illinois Press, 1980); G. Higgins, A.H. Kassam, L. Naiken, G. Fischer and M. Shah, "Potential Population Supporting Capacities of Lands in the Developing World," Technical Report of FAO, IIASA and UNFPA Project Int/75/P13, *Land Resources for Populations of the Future* (Rome: FAO, 1983); Ragnar Overby, "The Urban Economic Environmental Challenge: Improvement of Human Welfare by Building and Managing Urban Ecosystems," presented at POLMET 85, Urban Environmental Conference, Hong Kong, 1985; M.A. Harwell and T.C. Hutchinson, *Environmental Consequences of Nuclear War*, Vol.II, SCOPE 28 (Chichester, UK: John Wiley, 1986); *Action Plan Netherlands*, Friends of the Earth (Netherlands). A fascinating intellectual history of a part of this debate, with particular reference to Serhii Podolinski, Ludwig Boltzmann, Rudolf Clausius, Frederick Soddy, is provided by agro-economist Juan Martinez-Alier, *Ecological Economics: Energy, Environment, and Society* (Oxford: Basil Blackwell, 1987).

14. Peter M. Vitousek, Paul R. Ehrlich, Ann H. Ehrlich and Pamela A. Mateson, "Human Appropriation of the Products of Photosynthesis," *BioScience* Vol.34, No.6 (1986): 368-373; and D. Pauly and V. Christensen, "Primary Production Required to Sustain Global Fisheries," *Nature* (forthcoming) 1995.

15. William E. Rees, "Achieving Sustainability: Reform or Transformation?," *Journal of*

Planning Literature Vol.9, No.4 343=361 (1995); and "Revisiting Carrying Capacity: Area-based Indicators of Sustainability," *Population & Environment* (1995, in press).

16. We have previously defined this individual footprint as the "personal planetoid." (See William Rees and Mathis Wackernagel, "Ecological Footprints and Appropriated Carrying Capacity: Measuring the Natural Capital Requirements of the Human Economy," in *Investing in Natural Capital: The Ecological Economics Approach to Sustainability*, ed. A-M. Jansson, M. Hammer, C.Folke, and R. Costanza (Washington: Island Press, 1994).

3

FUN WITH FOOTPRINTS: METHODS & REAL-WORLD APPLICATIONS

If you would like to estimate the Ecological Footprint of projects, policies, programs or particular technologies, read this chapter. It describes our present approach to such calculations and gives examples of real-world applications.

Making the Ecological Footprint Idea Work

In theory, the Ecological Footprint (EF) of a population is estimated by calculating how much land and water area is required on a continuous basis to produce all the goods consumed, and to assimilate all the wastes generated, by that population. However, attempting to include all consumption items, waste types and ecosystem functions in the estimate would lead to intractable information and data-processing problems. We therefore use a simplified approach in our "real-world" research and in the examples to follow. In general, we:

- Base calculations on the assumption that the current industrial harvest practices (e.g., in agriculture and forestry) are sustainable, which they often are not.
- Include only the basic services of nature. As the assessments are refined, additional natural functions can be included. Human activities directly and indirectly appropriate nature's services through the harvest of renewable resources, extraction of non-renewable resources, waste absorption, paving over, fresh water withdrawal, soil contamination, and other forms of pollution (including ozone depletion). At this point, our research has concentrated on the first four activities.
- Try not to double-count when the same area of land provides two or more services simultaneously. For example, an area might be growing timber or pulp-wood while at the same time collecting water subsequently used for domestic purposes or irrigation. In this case, only timber production — the larger land area — would be included in the Footprint estimate.
- Use a simple taxonomy of ecological productivity involving eight land

61

(or ecosystem) categories.

- Are only beginning to include marine areas. Although humans already use critical marine ecosystems as intensely as the land, the sea provides a small fraction of human consumption and is less subject to policy and management manipulation than are terrestrial ecosystems (see Box 3.1).

Because of the first and second of these simplifications, our results present a conservative picture of humanity's demands on land. For example, assuming that present land uses are sustainable greatly underestimates the area of land required for truly sustainable production. High-input production agriculture typically depletes cropland soils in North America 10 to 20 times faster than they can regenerate. In other words, to compensate for soil loss, land in crop production should be left fallow for a decade or more for each year of cultivation. If we accounted for this regeneration period in our analyses, it would increase the area appropriated for crops by a factor of at least 10. Similarly, current forestry practices may not be sustainable: it is questionable whether, with current harvest practices, planned 70-year rotation periods can be sustained for more than two to three harvests. In addition, assumed yields can be maintained only if productivity is not reduced by pests or fires.[1]

We call the ratio of the land area that would be required under sustainable land-use and harvest practices to the land area that is actually required using prevailing production methods the "sustainability factor." (The sustainability factor is 10 to 20 in the agriculture example.) The magnitude of this factor is proportional to the rate of natural capital depletion and indicative of our present reliance on, and confidence in, technology (often itself based on non-renewable resources) to maintain long-term productivity. In this sense, our Footprint estimates could be challenged as excessively optimistic. They grossly underestimate the land requirement of the economy as it would be, unsubsidized by natural capital depletion and technological inputs. Indeed, a technological pessimist would be justified in multiplying components of our EF calculations by their corresponding sustainability factors, greatly increasing the aggregate area.

Our simplified approach might also be criticized for not considering a larger variety of biophysical life-support services, particularly those that are not directly associated with land-based renewable resource production. While the scope of the present analyses is restricted, we do not think this limitation weakens the conceptual or consciousness-raising value of EF analysis for several reasons. *First*, there is virtue in accurate simplicity. However complete a theory or model purports to be, it cannot include all aspects of reality. By definition, every model is necessarily an abstraction from, and interpretation of, a more complex reality. To capture the essence of the thing it represents, a model must incorporate those key variables and limiting factors which determine and explain the behavior of that real-world entity. In short, good theory finds a balance between complexity and simplicity — to be effective in guiding policy, models must be good enough to capture the essence of reality but simple enough to be understood and applied. For example, the human body temperature is a good indicator of human health. The theory that says "temperatures much over 37° Celsius are bad" is an enormous simplification, but a highly operational one — i.e., the theory is in most cases "good enough" to indicate illness. Similarly, EF analyses need not include all consumption items, waste categories and ecosphere functions to have diagnostic value.

Consistent with this approach, models concerned with the biophysical dimensions of sustainability should concentrate on understanding potentially limiting factors. Current trends suggest that the factors most likely to impose limits on human activity are certain forms of natural capital and the life-support functions they perform. In the 1970s, the limits-to-growth debate was largely concerned about the depletion of non-renewable resources such as metal ores and fossil fuel. In contrast (and ironically), a more likely bottleneck today seems to be the declining stocks of *renewable* natural capital such as fish, forests, soil and clean water. EF analysis therefore focuses on the renewable natural capital requirements of the economy and recognizes nature's capacity for self-renewal as a major limiting factor. Non-renewable resources are presently included in the Footprint only through the impacts of extraction and processing energy use and the direct occupation of land by mining infrastructure. More detailed analyses would also account for pollution effects. How we translate fossil fuel use into land equivalents is discussed below.

A *second* reason for keeping things simple is that certain ecosystem functions are analytically intractable. For example, it is difficult to quantify the connection between such generalized life-support services as global heat distribution, biodiversity and climate stability and either *per capita* demand for these services or associated ecosystem area. While these life-support services are essential for well-being and we all "consume" them, they cannot as yet be incorporated directly into the Ecological Footprint.

Calculation Procedure

As previously explained, the EF concept is based on the idea that for every item of material or energy consumption, a certain amount of land in one or more ecosystem categories is required to provide the consumption-related resource flows and waste sinks. Thus, to determine the total land area required to support a particular pattern of consumption, the land-use implications of each significant consumption category must be estimated. Since it is not feasible to assess land requirements for the provision, maintenance and disposal of each of the tens of thousands of consumer goods, the calculations are confined to select major categories and individual items.

Estimating the Ecological Footprint of a defined population is a multi-stage process. The basic structure of our approach is as follows. While the description refers to resource consumption, the same logic would apply to many categories of waste production and assimilation:

First we estimate the average person's annual consumption of particular items from aggregate regional or national data by dividing total consumption by population size. This is much simpler than attempting to estimate individual or household consumption by direct measurement! Much of the data

---

BOX 3.1: The Human Footprint in the Sea [2]

We have so far not included the marine area appropriated for human use in present Footprint estimates for several reasons. First, despite their vast area, the world's oceans provide only a small fraction of human direct consumption; second, despite this small contribution, the seas are already over-exploited by humans; third, there seems to be less scope for management manipulation of the seascape than of the landscape; forth, and most important, inclusion of the sea is generally not necessary for Footprint analysis to "make the case" that the total human load exceeds global carrying capacity. That said, ongoing work does include the marine area associated with seafood consumption to facilitate international comparisons and for possible incorporation into extended Footprint analysis. These studies support the findings of terrestrial Footprint analyses. Some of the factors we are taking into account are as follows:

Wild fish stocks, the dominant renewable resource from freshwater and marine ecosystems, provide less than two-and-a-half percent of the human food requirements as measured by nutritional energy content. This corresponds to about 16 percent of world consumption of animal protein. At the same time, it is unlikely that the resource yield from oceans, lakes and rivers can be much expanded economically; most fisheries are already over-harvested as humankind has become the dominant top carnivore in the sea. Indeed, the United Nations Food and Agriculture Organization (FAO) estimates that the global harvest of marine food approaches 90 percent of the theoretical maximum yield of the desirable species, if it has not reached it already. In fact, "...the *per capita* seafood supply, which peaked at 19 kilograms in 1989 and has since fallen, will be back down to 11 kilograms..." by 2030, according to Lester Brown from the Worldwatch Institute.

Some might argue that this scarcity could be overcome through fish-farming. However, fish-farming only shifts the ecological demand to other ecosystems such as the terrestrial cropland necessary to produce the feedstock for the fish farms or the water area to produce the algae that is fed to fish in the form of pellets. In fact, according to Carl Folke from the Beijer Institute in Stockholm, intensive salmon farming requires solar fixation by plankton from a sea surface area that is about 50,000 larger than the surface area covered by the farm cages.

needed for preliminary assessments is readily available from national statistical tables on, for example, energy, food, or forest products production and consumption. For many categories, national statistics provide both production and trade figures from which trade-corrected consumption can be assessed:

trade-corrected consumption = production + imports - exports

The next step is to estimate the land area appropriated *per capita* (aa) for the production of each major consumption item 'i.' We do this by dividing average annual consumption of that item as calculated above ['c,' in kg/capita] by its

---

It could also be argued, of course, that the oceans are used extensively as a dumping ground for waste and should be included in Footprint analysis on this basis. However, because ocean currents and upwellings produce significant material and heat exchange among all the seas of the world, the large and unknown dilution factor makes it difficult to translate waste discharges at sea into a well-defined appropriated area. In any event, bioaccumulation of toxic contaminants in food chains often renders measurements of ambient concentrations ecologically meaningless. On the other hand, because non-degradable toxic organic wastes (such as DDT and PCBs) and non-organic waste (such as heavy metals or radioactive substances) do accumulate in ecosystems, this can be reflected in EF analysis to the extent that heavily contaminated areas become unavailable for human consumption. Such contamination reduces the local "carrying capacity" available to human beings and expands the Footprint into alternative productive areas on land or sea.

For those interested in that part of the human marine Footprint associated with seafood consumption, a generalized first approximation can be calculated as follows. We start by dividing the fish catch by total productive ocean area. The maximum sustainable yield of the oceans is about 100 million tonnes of fish per year. While the seas occupy about 71 percent of the Earth's total surface (about 362 million square kilometres), less than 8.2 percent of this (or about 29.7 million square kilometres) is responsible for about 96 percent of the global fish-catch. In other words, average annual production is about 33.1 kg of fish per productive hectare or 0.03 hectares per kilogram of fish. An equal "seashare" of ocean (productive area divided by total human population) would therefore be about 0.51 hectares *per capita*, which corresponds to about 16.6 kilograms of fish per year. For comparison, Japan, one of the great fishing nations, accounts for about 12 percent of the global catch and her people consume 92 kg of fish *per capita* annually. This is about 5.4 times the estimated global maximum sustainable yield *per capita*, giving the average Japanese a marine EF approaching 2.8 ha. Clearly the whole world cannot aspire to Japan's level of seafood consumption.

Similar calculations could also be performed for freshwater fisheries.

average annual productivity or yield ['p,' in kg/ha]:

$$aa_i = c_i / p_i$$

Of course, many consumption items (e.g., clothing and furniture) "embody" several inputs and we have found it useful to estimate the areas appropriated by each significant input separately. Ecological footprint calculations are therefore both more complicated and more interesting than appears from the basic concept.

We then compute the total ecological footprint of the average person ('ef') — i.e., the *per capita* footprint — by summing all the ecosystem areas appropriated (aa_i) by all purchased items (n) in his or her annual shopping basket of consumption goods and services:

$$ef = \sum_{i=1 \text{ to } n} aa_i$$

Finally we obtain the ecological footprint (EF_P) of the study population by multiplying the average *per capita* footprint by population size (N):

$$EF_P = N(ef)$$

In some cases where the total area used is available from national statistics, we compute the *per capita* footprint by dividing by population.

Most of our footprint estimates are based on average national consumption and world average land yields. This is a standardization procedure that facilitates "general case" comparisons among regions or countries. (It is also fairly realistic for many countries given the increasing reliance on multi-lateral trade flows and appropriations from the global commons.) However, for more sophisticated or detailed analyses, it may be necessary or desirable to base the Footprint estimate on regional or local consumption and productivity statistics. With sufficient data, locally accurate EFs of consumer units as small as specific municipalities, households, and even individuals can be estimated. For example, we have sometimes found it interesting to compare the Ecological Footprint estimated from locally specific data to the "first approximation" based on national average consumption and global productivities. Such comparisons reveal the effects of regional variation in consumption patterns, productivities, and management approaches on the size of the local Footprint. They can also help identify and eliminate data gaps, errors, and apparent contradictions in the calculations.

Consumption categories

To simplify data collection, we have generally adopted data classifications used for official statistics. On this basis, we have found it useful to separate

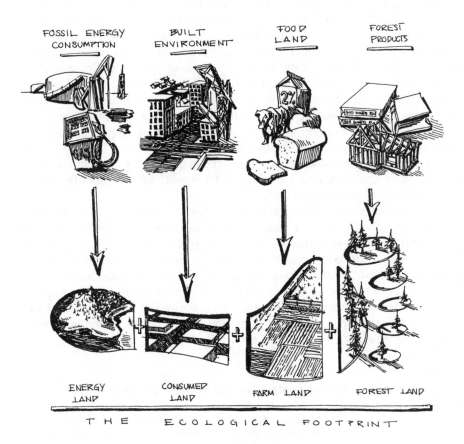

Figure 3.1: Converting Consumption into Land Area.
The production and use of any good and service depends on various types of
ecological productivity. These ecological productivities can be converted to land-
area equivalents. Summing the land requirements for all significant categories of
consumption and waste estimates the EF for the reference population.

consumption into five major categories:
1. food
2. housing
3. transportation
4. consumer goods
5. services.

For more refined analyses, these categories can be subdivided as required. For example, the food component of the Footprint could be "assembled" by considering vegetable- and animal-based products separately. Transportation could be separated into public and private transportation. Such subcategories should be defined strategically in order to answer specific policy questions about that item. For each consumption item, a detailed analysis would encompass all the embodied resources that go into the production, use and disposal of that item. The "embodied" energy and resources of a commodity refers to the total quantities of energy and material that are used during the life cycle of that commodity for its manufacture, transport and disposal. "Energy intensity" refers to the embodied energy per unit of a good or service. Similarly, we can speak of the "embodied Ecological Footprint" of a commodity as its contribution to the consumer's EF.

These principles and definitions hold true for both goods and services, even though "services" are often considered to be essentially "non-material." The fact is that services are also sustained by energy and material flows. Even the transmittal of information requires both energy and physical carriers such as paper or wires and, to make it accessible, people need material interfaces such as screens or radios. Banks may produce nothing material but all their operations from money transactions, through the computerized generation of bank statements, to the construction and operation of bank buildings and infrastructure consume physical energy and resources.

Table 3.1:
The 8 main land & land-use categories for Footprint assessments

I) energy land:	a. land "appropriated" by fossil energy use	(ENERGY OR CO_2 LAND) **Note:** If we opt for fuel crops,this would remove some land from categories c, d, e or f.
II) consumed land:	b. built environment	(DEGRADED LAND)
III) currently used land:	c. gardens	(REVERSIBLY BUILT ENVIRONMENT)
	d. crop land	(CULTIVATED SYSTEMS)
	e. pasture f. managed forest	(MODIFIED SYSTEMS)
IV) land of limited availability:	g. untouched forests	(PRODUCTIVE NATURAL ECOSYSTEMS)
	h. non-productive areas	(DESERTS, ICECAPS)

Table 3.2: Productivity of various energy sources.
The energy Footprint varies inversely as the productivity of an energy
source: the higher the productivity, the smaller the Footprint.

Energy Source	Productivity [in Gigajoules per hectare per year].	Footprint for 100 Gigajoules per year [in hectares]
Fossil fuel		
ethanol approach	80	1.25
CO_2 absorption approach	100	1.0
biomass replacement		
approach	80	1.25
Hydro-electricity (average)	1,000	0.1
lower course	150-500	0.2-0.67
high altitude	15,000	0.0067
Solar hot-water	up to 40,000	0.0025
Photovoltaics	1,000	0.1
Wind energy	12,500	0.008

Numerous sources can be used to quantify direct consumption and associated embodied resources. Statistics on waste streams, household and national expenditure, metabolic rates, diets, trade and resource flows can be consulted — and checked, one against the other (see Box 3.2).

Land and land-use categories

Our EF calculations are based on the following eight major land categories (Table 3.1). This classification is similar to that used by The World Conservation Union (IUCN).[3]

The "energy land" component of the EF can be computed in several ways (see below). Some methods estimate the area that would be required to grow fuel crops to replace our depleting stocks of fossil energy. If this notion seems far-fetched, keep in mind that fossil fuels are the product of ancient photosynthesis and the accumulation of biomass in forests and swamps that grew over much of the Earth's surface millions of years ago. William Catton therefore refers to these lands as "phantom land." The ecosystems are long gone but, in effect, we are still using them — or at least their productivity — today.[4] Catton points out that humanity is using this former productivity thousands of times faster than it accumulated and that nature is not able to replace it. In the absence of contemporary managed terrestrial carbon sink reserves, we are imposing a burden on future generations: less carbon-based energy stocks and elevated levels of atmospheric CO_2. In other words, we are using two kinds of natural income and liquidating critical natural capital without replacement or compensation.

Not all categories of ecologically productive land are equally accessible or directly harvestable by humans. Certainly, given growing concerns about climate change, we should approach category 'g' with great caution. This category represents virgin forest ecosystems whose harvest would lead to a massive net CO_2 release that would be recovered only after 200 years of subsequent ecological production on this land.[5] Some of these forest lands are still accumulating carbon and also serve as biodiversity refuges that should not be disturbed. Land in category 'h' includes deserts and ice-fields such as the Sahara and Antarctica and is regarded as ecologically unproductive for human purposes.

The remaining land categories provide a variety of goods and services (natural income) in support of human activities, from the provision of commercial energy, through space for cities and the absorption of waste, to the preservation of biodiversity. Here is how we convert these services to their land area equivalents for EF analysis.

BOX 3.2: Data Sources for Ecological Footprint Analyses

There are many sources of data for Footprint analyses. For approximate comparisons, a compendium such as the biannual report of the World Resources Institute may be sufficient. However, international statistics often focus mainly on production and trade, omitting consumption, and are often in dollars (rather than biophysical units), which decreases their usefulness. The list below maps the diversity of possible data sources that can be used for Footprint calculations. Please let us know if you come across good additional sources!

Global and National Statistics

» Food and Agriculture Organization of the United Nations or FAO (*The State of Food and Agriculture; FAO Yearbook: Trade; FAO Yearbook: Production*, all annual)
» International Road Transportation Union or IRTU (*World Transportation Data*, annual)
» United Nations Development Program or UNDP (*Human Development Report*, annual)
» The World Bank (*World Development Report*, annual)
» World Resources Institute or WRI (*World Resources*, biannual, also available on computer disk)
» Worldwatch Institute (*State of the World*, *Vital Signs*, both annual, the latter also available on computer disk)
» United Nations statistics
» Government publications with national statistics on:
    - Consumption, Economic Production and Trade
    - State of the Environment

i) Land requirements for commercial energy. This section discusses the land "use" implications of consuming fossil fuel, hydroelectricity and other renewable energy sources (Table 3.2).

Most of the energy on which human life depends comes from the sun. In fact, life on Earth is powered by a solar flux of about 175,000 terawatts. One terawatt is one trillion (or 1,000,000,000,000) watts or joules per second. This is the same energy required to lift one million tonnes 100 metres every second. In comparison, a standard light bulb radiates 60 watts of heat and light.

The commercial energy flow through the human economy amounts to "only" 10 terawatts. However, if we had to produce these 10 terawatts of commercial energy using contemporary photosynthesis we would need an enormous area of land: of the 175,000 solar terawatts, fewer than 150 are transformed into plant biomass by photosynthesis ("Net Primary Productivity"). Only a small fraction of this can be harvested and still a smaller fraction converted to useful fuel.

In the following discussion, the *energy-to-land* ratio describes how much

---

- Transportation
- Land-use
- Housing
- Energy
- Agriculture and Forestry

Reference and Handbooks:

» Engineering, ecology, resource management and agricultural handbooks
» Professional handbooks on topics such as agriculture, biological resources, energetics, chemistry, etc.
» Handbooks on energetics and life cycle analysis
» Handbooks on ecological cycles and biological productivity (e.g., carbon cycle, Net Primary Productivity)
» Transportation handbooks
» Engineering handbooks on the energy aspects of housing, transportation, chemical processes, technological efficiency, etc.
» Household ecology guides
» Encyclopedias, yearbooks and almanacs
» Cookbooks (for nutritional value of food, cooking energy, etc.)

Research Papers

» Reports in the popular and scientific press on consumption, energy efficiency, ecological productivity, etc.
» Special issues reports by Non-government Organizations (NGOs), government agencies, institutes (e.g., Greenpeace reports on cars, paper consumption).

commercial energy per year could be provided by one hectare of ecologically productive land. The units used are gigajoules per hectare per year, or GJ/ha/yr. One gigajoule stands for one billion joules; 1,000 gigajoules per second is equal to one terawatt.

We have used three approaches to converting **fossil energy** consumption into a corresponding land area. Each is based on a different rationale, but all produce approximately the same results — the consumption of 80 to 100 gigajoules of fossil fuel per year corresponds to the use of one hectare of ecologically productive land.

The *first method* calculates the land required to produce a contemporary biologically-produced substitute for liquid fossil fuel. In effect, this is the area of land needed to bring Catton's "phantom land" back to life. This approach reasons that a sustainable economy requires a sustainable energy supply, i.e., it should not be dependent on depletable fossil capital. Moreover, if the fuel is carbon-based it is preferable to use carbon that is already cycling actively in the ecosphere rather than carbon that has been stored for millennia in an inactive pool. This approach avoids further CO_2 accumulation in the atmosphere.

Ethanol is one such potentially renewable energy carrier that is technically and qualitatively equivalent to fossil fuel. It is a homogeneous, concentrated fuel that can easily be stored and transported, and can power human processes the same way fossil hydrocarbons do. For these reasons it is already being used in some places as a supplement to gasoline. The land area corresponding to fossil fuel consumption can therefore be represented as the productive land necessary to produce the equivalent amount of ethanol. This area comprises land needed to grow the plant material (biomass) for both the ethanol fuel and the necessary processing energy. The most optimistic estimates for ethanol productivity suggest a net productivity of 80 gigajoules per year per hectare of ecologically productive land. [6]

Methanol is another possible substitute for fossil fuel. Calculations suggest that each kilogram of wood distilled would yield 10.5 to 13.5 megajoules of methanol. (One megajoule corresponds to one million joules or one thousandth of a gigajoule.) New Zealand tree plantations, at 12 tonnes of wood per hectare per year, are among the world's most productive "forests" and would yield a land-for-energy ratio of 120 to 150 gigajoules per hectare per year. However, the productivities typical of Canadian, Russian or Scandinavian, forests would yield only 17–30 gigajoules per hectare per year (approximately 55–68 in the U.S.).[7]

The *second method* estimates the land area needed today to sequester the CO_2 emitted from burning fossil fuel. The argument for this approach is that fossil carbon (in the form of CO_2) cannot be allowed to accumulate in the atmosphere if we wish to avoid possible climate change. If we continue to consume excessive quantities of fossil fuel we have a responsibility to manage its waste products. This approach requires that we calculate the amount of "carbon sink" land required to assimilate the fossil CO_2 that we are injecting into the atmosphere.

Forest ecosystems and peat bogs are among those natural systems that can be significant net assimilators of CO_2. Young to middle-aged forests accumulate CO_2 at the highest rate over a 50- to 80-year time span. Data on typical forest productivities of temperate, boreal and tropical forests show that average forests can accumulate approximately 1.8 tonnes of carbon per hectare per year.[8] This means that one hectare of average forest can sequester annually the CO_2 emission generated by the consumption of 100 gigajoules of fossil fuel.

The *third method* of converting fossil energy use into a corresponding land area estimates the land area required to rebuild natural capital at the same rate as fossil fuel is being consumed. This builds on an argument put forward by World Bank economist Salah El Serafy.[9] If we accept

that a society is not sustainable if its economy depends on the depletion of real wealth (natural capital), then any society using non-renewable resources should invest a portion of the revenues so generated in building up an equivalent value of manufactured capital or renewable resource assets. This approach — replacing what is consumed — addresses directly the constant capital stocks criterion for sustainability, which recognizes that equity between generations is a precondition for sustainability. Calculations show that one hectare of average forest could accumulate about 80 gigajoules of recoverable biomass energy per year in the standing timber. In other words, if we assume that depleted natural capital must be replaced, the land-for-energy ratio amounts to 80 gigajoules of biomass energy per hectare per year. (Once economic reserves of fossil fuels are used up and we start cropping the energy land, this method converges with the first.)

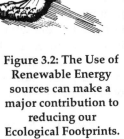

Figure 3.2: The Use of Renewable Energy sources can make a major contribution to reducing our Ecological Footprints.

The CO_2 assimilation method results in the smallest EF attributable to fossil fuel consumption. Many reviewers felt that this approach would enjoy the highest public acceptance. It implies no radical shift from fossil fuels yet accepts the need to stop greenhouse gas accumulation. Therefore, we chose **one hectare per 1.8 tonnes of carbon emitted each year (one hectare per 100 gigajoules per year)** from the CO_2 method as the land-for-energy ratio for fossil fuel. We use this ratio in all current EF assessments.

Note that if electricity is generated from fossil fuel with a typical efficiency of 30 percent, the EF per unit of end-use energy is over three times larger than if the fossil fuel were used directly.

Renewable energy sources provide much higher productivities (smaller EFs) than fossil fuel. For **hydroelectricity,** the area requirements can be estimated by dividing the flooded land behind dams, plus the land area occupied by high voltage power line corridors, by its annual electricity production. University of Manitoba Geographer Vaclav Smil suggests hydroelectricity productivities of 160 to 480 gigajoules per hectare per year for lower-course dams (in the 50 to 200 megawatt size), 1,500 to 5,000 gigajoules per hectare per year for middle- and upper-course dams, and 15,000 gigajoules per hectare per year for alpine high-altitude dams. Similarly, Michael Narodoslawsky and his colleagues at the Technical University of Graz, Austria estimate the productivity of typical hydro-power stations at about 1,500 gigajoules per hectare per year (not including the space requirements of power lines). Including power-lines would reduce this ratio to approximately 1,000 gigajoules per hectare per year. In contrast, David Pimentel and his team from Cornell University calculate an average hydroelectric productivity of only 47 gigajoules per hectare per year for the U.S., ranging from 4.5 gigajoules per hectare per year for lower-course systems up to 7,300 gigajoules/ha/yr. for high-altitude dams. (These latter data suggest that hydro plants that would yield less than 100 gigajoules per hectare per year — typical for biofuel — might be ecologically inefficient, particularly as dams in the lowlands flood areas of high ecological productivity.)[10] All these data indicate that a land-for-energy ratio of one hectare for each 1,000 gigajoules of continuous generating capacity would not be unreasonable for general EF calculations. (Note that this still does not account for other negative ecological effects such as impact on fisheries.) These Footprint areas would fall into the built environment category. However, when corridor land is made available for pasture, care should be taken to avoid double counting.

To date we have not included hydroelectricity consumption in our EF calculations. However, a preliminary estimate for Canada yields the following: according to the World Resources Institute, in 1991 Canada produced 1,111 petajoules (or 1,111 million gigajoules) of hydroelectricity.[11] At a land-for-energy ratio of one hectare per 1,000 gigajoules per year, this would add another 0.04 hectares to the average Canadian's Footprint for flooded land and transmission lines.[12]

Other forms of renewable energy reach quite impressive yields. Preliminary analysis suggests that large-scale **photovoltaic electricity** might produce

100 to 1,000 gigajoules per hectare per year, confirmed by the experience of a 2 hectare photovoltaic plant in the Swiss Alps, which delivered in its first year of operation about 1,000 gigajoules of electricity per hectare to the power grid.[13] Other examples of renewable energy production include wind generation in America's windiest places, which might score between 250 and 500 gigajoules per hectare per year. If we consider that the physical footprints of windmills occupy only two percent of the wind-farm area, allowing some other functions on the land, the productivity of the windmill rises to 12,500 to 25,000 gigajoules per hectare per year. Well-designed low-temperature solar collectors (for domestic hot water applications) can achieve 10,000 to 40,000 gigajoules per hectare per year.

It is important to recognize not only that in many areas the use of renewable energy sources such as photovoltaic cells, windmills and hot water solar collectors would significantly reduce the fossil fuel components of our present EFs, but also that these sources do not themselves require any direct use of ecologically productive land.

We do not incorporate **nuclear energy** in current EF assessments. On the surface, nuclear energy needs little space. In fact, including the complete fuel cycle of mining, processing of uranium ores, uranium enrichment, production of fuel elements, reprocessing of spent fuel, and storage of radioactive wastes, and *assuming no accidents*, each hectare occupied produces over 50,000 gigajoules per year. In other words, the productivity of well-functioning nuclear power plants seems to exceed that of the most efficient ethanol technology by two to three orders of magnitude. However, if we consider the impact of accidents — lost bioproductivity and contaminated land — the tables turn. In the case of Chernobyl, we estimate that energy productivity decreased to less than 20 gigajoules per hectare for the years immediately following the accident. In any event, the shattered popular trust in nuclear safety, the fact that peaceful use and military applications are interwoven, and the seemingly unsolvable problem of radioactive waste — which becomes an irresponsible burden for future generations — suggest that nuclear power is not a viable energy option today.

ii) Accounting for built-up land. Paved-over, built upon, badly eroded or otherwise degraded land is considered to have been "consumed" since it is no longer biologically productive. This means that total future bioproductivity has been reduced. As demand increases, it may become necessary to upgrade inferior land elsewhere to compensate for this lost productivity.[14] An additional debit would then be charged against the degraded land account for the energy, material and time expended to restore its productivity. (Economists generally overlook the fact that the substitution of human-made capital and labor for depleted natural capital and its functions carries an opportunity cost in the form of reduced economic productivity — the necessary expenditures are not available for other forms of investment or consumption.)

iii) Provision of water. In many regions of the world, the consumption of fresh water for human use compromises other possible use of this water or of the land required to "collect" it. In addition, energy and material is consumed in transporting the water. Thus, depending on the source of the water, EF analysis should account for the opportunity cost of water withdrawal and the energy costs of transporting the water. (The additional land needed to compensate for lost ecological productivity at the source may show up in the agricultural [crop- and pastureland] accounts, for example.) Catchment areas for water should also be included to the extent that water collection can be separated from other bioeconomic functions of the catchment area (otherwise, it would lead to double counting). In drier areas, these catchment areas can be of substantial size. For example, in Australia, for every city dweller about 0.27 to 0.37 hectares of land are set aside for water collection.[15]

iv) Absorption of waste products. Nature's capacity to absorb human-made waste is finite. However, substantial flows of nutrients and domestic organic wastes, if adequately distributed, can be broken down and the by-products recycled by local ecosystems with little exclusive addition to the EF. (Only the land required for pre-release sewage treatment facilities need be included. Nature's final processing of the residuals takes place in waters or on lands used and counted in the EF for other purposes.) On the other hand, what cannot be degraded and assimilated accumulates locally, or is carried away by water and air only to accumulate elsewhere, in the sea, or in global food chains. Contamination of soil, water, and airsheds may reduce productivity or contaminate the products of nature to the extent that they become unfit for human consumption. Where significant, these land and productivity losses should become part of the waste disposal Footprint. Similarly, to the extent that depletion of the atmospheric ozone layer eventually reduces bioproductivity (through damage to photosynthesis by increased UV_B radiation), this loss should be added to the EF area. In our EF examples to date, we have not accounted for waste absorption and pollution damage with the exception of the major contribution from CO_2 sequestering.

v) Protecting biodiversity. Biodiversity is threatened by the irreversible loss and fragmentation of wilderness areas on all continents. There is an ongoing debate about how much wilderness should be set aside, and in what configuration, to secure both adequate biodiversity and global ecological stability. Ecologist Eugene Odum has suggested that a third of every ecosystem type should be preserved to secure biodiversity. The Brundtland Commission proposed, seemingly arbitrarily, that at least 12 percent of the Earth's land area (or about 2 billion hectares) should be set aside for this task. The fact is, we have little idea how much natural habitat is required for the survival of other species, let alone to ensure our own ecological security. To what extent do modified and heavily exploited ecosystems such as well-managed forests conserve biodiversity and provide basic life-support functions? As noted, land category h refers to the about 1.5 billion hectares of nearly untouched forest ecosystems that both serve as a substantial carbon pool and provide habitat to the bulk of the Earth's species.[16] These 1.5 billion hectares correspond to just 9 percent of the Earth's terrestrial area, only one third of which is under protection; given current uncertainties and the scale of the potential hazard, ordinary prudence and the precautionary principle argue that this area should be left intact for the sake of global security.

The consumption — land-use matrix

Once the main consumption and land-use categories are defined, the connection between each consumption category and its land requirements must be established using the calculation procedure described above. The data are then assembled in a matrix that links consumption (rows) with land-uses (columns) (Table 3.3). Each of the data cells in the matrix represents a particular consumption item in terms of its corresponding "appropriated" land area.

The rows are divided into our five categories of consumption: food, housing, transportation, consumer goods and miscellaneous services. Note that the data for each category reflect not only the space directly occupied by individual consumption items (where relevant) but also the land "consumed" in producing and maintaining them. In effect, this becomes a life cycle analysis of the land implications of consumption. The housing category, for example, encompasses the land on which the house stands (including a proportionate share of urban land occupied by infrastructure), the land necessary to grow lumber for the house (or, alternatively, the energy land associated with producing bricks), and the energy land appropriated for space heating.

As in Table 3.1, the columns of the matrix are identified with the letters **A** to **F**, each representing one type of land-use. Column **A** represents the fossil energy land-equivalent for each consumption item using a land-for-energy ratio of one hectare per 100 gigajoules per year. Column **B** indicates the amount of built-over and degraded land. Column **C** shows garden-land, the area used mainly for vegetable and fruit production. (Typically, this land has the highest ecological productivity.) Column **D** contains other cropland, and column **E** the pastureland used for dairy, meat and wool production. Finally, column **F** includes the land committed to providing forest products. The **TOTAL** column shows aggregate land "occupancy" by each consumption category.

The Footprint data in Table 3.3 are based on global average ecological productivities. As noted above, this provides a reasonable approximation for several reasons. *First*, it reflects the increasingly diffuse real-world relationship between local consumption and corresponding global production. Many industrial urban communities depend little on local ecological productivity —

the ingredients of most of their consumption items typically originate in distant regions all over the world. *Second*, having a globally-adjusted measurement unit enables easy international comparisons of consumption impacts. *Third*, it facilitates accounting while not distorting the aggregates. Thus, if for some reason we wished to compare a given population's EF computed on the basis of global average productivity with the EF it might have based on the quality of locally available land, productivity adjustments to land area must be made. For example, if agricultural land in a particular region is twice as productive as the world average, a hectare of the local land would correspond to two hectares of average land and the EF based on local productivity would shrink accordingly. Of course, the sum of all such regionally-adjusted land areas would be equivalent to the globally available productive land area.

To reiterate, the calculation procedure described is conceptually simple and easy to perform. While an Ecological Footprint analysis could be done from scratch with detailed data on an individual's or a community's consumption patterns, we generally begin with aggregate (e.g., national or provincial/state) data. The analysis can later be elaborated with more detailed data on specific communities, regions, or even individual technologies, as necessary or useful.

The strength of the EF analysis is its ability to communicate simply and graphically the general nature and magnitude of the biophysical "connected-ness" between humankind and the ecosphere. In a single index, the Ecological Footprint captures the essence of humankind-nature relationships as mani-fested through consumption. As explained in the preceding chapter, EF calculations are static. They provide an ecological snapshot of economy-land relationships at a particular point in time. However, historical trends can be captured by reconstructing the EF for a series of such points. It thus provides a starting point either for more detailed analysis of specific problem areas or for discussion of the broad policy implications for sustainable development.

The Ecological Footprint approach is sometimes wrongly criticized for disregarding the effects of technological improvements. The argument is that the EF of a population could be reduced if technology is able to substitute for certain resources or if efficiency gains enable us to enjoy equivalent or higher material standards with fewer resources. Either improvement could poten-tially decrease aggregate material consumption. Indeed, it is sometimes argued that massive efficiency gains could effectively "decouple" growth in *per capita* GDP from nature. (There is a growing literature on "eco-efficiency." However, see Box 4.1 for some of the counter-intuitive effects of efficiency strategies.)

It is true that EF analysis does not produce a dynamic picture of changing conditions. However, far from ignoring technology, EF analysis allows us to compare current ecological requirements and constraints with those that would result if specified technological improvements were widely imple-mented. For example, it would graphically reveal the effects on carrying capacity of a significant shift from fossil fuels to solar energy. And, through the use of time series, EF analysis can even provide a dynamic picture of changing conditions. Indeed, by showing the dependence of the economy on natural capital/income under any specified set of conditions, EF analysis provides an incentive to improve, an estimate of how far we have to go to achieve sustainability (the "sustainability gap"), and a yardstick to monitor the economy's progress toward reducing its load on nature. The latter could be achieved through technological decoupling *or* changing values — either would result in decreased material consumption.

The Footprint in Action: Adapting the Calculation Procedure to Specific Applications

After theory comes action. This section shows how the EF concept is applied using real data: we derive a detailed estimate of the average Canadian's Footprint and describe 16 other applications more briefly. In order not to overload the reader with numbers and statistics, we provide only summary results of these applications.

Since EF analysis can be applied at various scales (individual, household, region, nation, world), the first task is to define the population or economy whose appropriated carrying capacity we wish to estimate. We should keep in

Figure 3.3: Figuring Out Footprints is Fun. With a pocket calculator and a few statistical books such as *World Resources* we are ready to calculate some simple Footprint examples.

mind, however, that the basic EF results are most interesting and useful in comparative analyses. For example, we might wish to contrast a given population's Ecological Footprint with the land area that is actually available in that population's home region, or with the hypothetical Ecological Footprints that might result from changes in the population's lifestyle. How we intend to use the analysis will affect our data requirements. Let's get started.

1) How big is the Ecological Footprint of the average North American?

"So, how big *are* people's Footprints?" This is one of the first questions people ask when introduced to the EF concept. As noted in Chapter 1, the answer depends on such factors as income, personal values and behavior,

---

BOX 3.3: Some Examples:
Translating Consumption into Land Areas

Example 1: fossil energy consumption and carbon sinks

Question: How much ecologically productive land (i.e., carbon sink forest) would be required to sequester all the CO_2 released by the average Canadian's consumption of fossil energy? (See "total" in Column **A** of the consumption – land-use matrix [Table 3.3].)

The World Resources Institute reports that Canada's total commercial energy consumption was 8,779 petajoules (PJ or million gigajoules) in 1991. Of this amount, 926 PJ were generated by nuclear power and 1,111 PJ by hydro-dams. Hence, the fossil fuel consumption was (8,779 - 926 - 1,111 =) 6,742 PJ. Therefore, each of the 27 million Canadians in 1991 would consume...

$$\frac{6{,}742{,}000{,}000 \ [\text{GJ/yr.}]}{27{,}000{,}000 \ [\text{Canadians}]} = 250 \ [\text{GJ/year}] \text{ of fossil fuel.}$$

However, Statistics Canada reports a figure of 234 GJ *per capita* per year. Wishing to err on the side of caution we use the Statscan data. With the land-for-energy conversion ratio for fossil fuel of 100 GJ/ha/yr., the land requirement for the average Canadian therefore comes to...

$$\frac{234 \ [\text{GJ/cap./yr.}]}{100 \ [\text{GJ/ha/yr.}]} = 2.34 \ [\text{ha/cap.}] \text{ for sequestering the } CO_2 \text{ released by this fossil fuel.}$$

- - - - - - - - - - - - - - - -

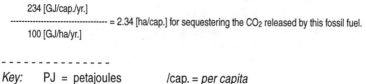

Key: PJ = petajoules   /cap. = *per capita*
         GJ = gigajoules   ha = hectares   t = tonne

consumption patterns, and the technologies used to produce consumer goods. There is, therefore, wide variation in Footprint size both among countries and individuals around the world. We can illustrate these points by summarizing our detailed calculations for the average Canadian's Ecological Footprint (Table 3.3) and contrasting the result with those for several other countries (Table 3.4). Note that while U.S. consumption patterns are roughly similar to Canada's *per capita* totals, their average Ecological Footprints are larger.

As noted, estimating the area of ecologically productive land needed to produce the natural resources and services used by an average Canadian involves several major steps: first we compile annualized statistics on five major categories of consumption and waste production and divide the totals

---

Example 2: productive forest area for paper

Question: How much forest area is dedicated to providing pulp-wood for the paper used by the average Canadian? (This corresponds to the cells "**f1**" (food wrappings), "**f40**" (packaging), "**f43**" (reading material) and the paper component of some of "**f2**" (household and construction paper) in the matrix of Table 3.3.)

Each Canadian consumes about 244 kilograms of paper every year. In addition to the recycled paper that enters the process, the production of each metric ton of paper in Canada currently requires 1.8 m^3 of wood. For Ecological Footprint analyses an average wood productivity of 2.3 [m^3/ha/yr.] is assumed. Therefore, the average Canadian requires...

$$\frac{244 \,[\text{kg/cap./yr.}] \times 1.8 \,[\text{m}^3/\text{t}]}{1,000 \,[\text{kg/t}] \times 2.3 \,[\text{m}^3/\text{ha/yr.}]} = 0.19 \,[\text{ha/capita}] \text{ of forest in continuous production for paper.}$$

Example 3: urban environment

Question: How large is the average Canadian's share of the nation's "built environment" (includes roads, residences, commercial and industrial areas, and parks — see "total" in Column 'b,' Table 3.3)?

The World Resources Institute reports 5,500,000 hectares of built-up land in Canada. Therefore, Canadians occupy...

$$\frac{5,500,000 \,[\text{ha}]}{27,000,000 \,[\text{Canadians}]} = 0.20 \,[\text{ha/capita}] \text{ of built-up land.}$$

Table 3.3: The consumption — land-use matrix for the average Canadian (1991 data)

Cell entries = ecologically productive land in [ha/capita]	A ENERGY	B DEGR.	C GARDEN	D CROP	E PASTURE	F FOREST	TOTAL
1 FOOD	**0.33**		**0.02**	**0.60**	**0.33**	**0.02**	**1.30**
11 fruit, vegs., grain	0.14		0.02	0.18		0.01?	
12 animal products	0.19			0.42	0.33	0.01?	
2 HOUSING	**0.41**	**0.08**	**0.002?**			**0.40**	**0.89**
21 constn./maint.	0.06					0.35	
22 operation	0.35					0.05	
3 TRANSPORTATION	**0.79**	**0.10**					**0.89**
31 motorized private	0.60						
32 motorized public	0.07						
33 transp'n of goods	0.12						
4 CONSUMER GOODS	**0.52**	**0.01**		**0.06**	**0.13**	**0.17**	**0.89**
40 packaging	0.10						
41 clothing	0.11			0.02	0.13	0.04	
42 furniture & appli.	0.06						
43 books/magazines	0.06					0.03?	
44 tobacco & alcohol	0.06			0.04		0.10	
45 personal care	0.03						
46 recreation equip.	0.10						
47 other goods	0.00						

5 SERVICES						0.30
51 gov't (+ military)	0.29	0.01				
52 education	0.06					
53 health care	0.08					
54 social services	0.08					
55 tourism	0.00					
56 entertainment	0.01					
57 bank/insurance	0.01					
58 other services	0.00					
	0.05					
TOTAL	2.34	0.20	0.02	0.66	0.46	0.59	4.27

(0.00 = less than 0.005 [ha] or 50 [m²]; blank = probably insignificant; ? = lacking data)

ABBREVIATIONS

a) **ENERGY** = fossil energy consumed expressed in the land area necessary to sequester the corresponding CO_2.

b) **DEGR.** = degraded land or built-up environment.

c) **GARDEN** = gardens for vegetable and fruit production.

d) **CROP** = crop land.

e) **PASTURE** = pastures for dairy, meat and wool production.

f) **FOREST** = prime forest area. An average roundwood harvest of 163 [m³/ha] every 70 years is assumed.

for items in these categories by total population to determine average levels. (Consumption includes: direct household consumption; indirect consumption such as the energy "embodied" in consumer goods; and consumption by businesses and government, which ultimately benefits the households. Services refers to schooling, policing, governance or health care.) Second, we convert these data on average consumption ("ecological load") into their corresponding land areas based on the ecological productivity of relevant ecosystem types. The average Canadian's Ecological Footprint is then obtained by summing the land requirements for the various consumption/waste categories. Since this area represents that portion of planetary productivity needed to support a single individual we sometimes refer to it as the average "personal planetoid." The results of these calculations are summarized in the consumption – land-use matrix shown in Table 3.3.

It seems that Canadians are formidable consumers! For example, on average, each Canadian eats about 3,450 kilocalories worth of food each day, 1,125 in the form of animal products. Most of this food is produced by energy-intensive agriculture and is highly processed before it reaches the dinner table. According to the World Resources Institute, Canadian settlements cover about 55,000 square kilometres — 0.2 ha *per capita* — and have been built mainly on agricultural land. On average, Canadians drive a car 18,000 kilometres per year, use approximately 200 kilograms of packaging, spend about $2,700 on consumer goods and another $2,000 on services. Energy and material consumption in Canada is typically four to five times the world average and, in most categories, the average American's consumption is even higher (see Table 3.4).[17]

Every year approximately 320 gigajoules of commercial energy are needed to power the average North American's activities, including the energy embodied in consumer goods and services. This is equivalent to the energy in 10 cubic metres of gasoline and, indeed, most of this energy is from fossil sources. The World Resources Institute reports that Americans use 287 gigajoules and Canadians 250 gigajoules of fossil energy *per capita* per year.[18] (Canada uses a larger percentage of hydroelectricity.) In Table 3.3 we account for only the fossil fuel part of the commercial energy consumption.

Government statistics provide a breakdown of energy consumption by economic sector. However, using these statistics directly for Footprint calculations may distort our picture of energy consumption at the household level because of the energy content of trade goods. The embodied energy in exports should not be included as domestic consumption while that in imports should be added in. Using this correction shows Canada, for example, to be a net exporter of embodied CO_2 emissions — and therefore of embodied energy — as a result of its international trade.[19] The Ecological Footprint applications described here are corrected for import-export balances only for the primary products of the forestry, agriculture and commercial energy sectors. For all

Table 3.4: Comparing people's average consumption in the US, Canada, India and the world[17]

Consumption per person in 1991	Canada	USA	India	World
CO_2 emission [in tonnes per yr]	15.2	19.5	0.81	4.2
Purchasing Power [in $ US]	19,320	22,130	1,150	3,800
Vehicles per 100 persons	46	57	0.2	10
Paper consumption [in kilograms/yr]	247	317	2	44
Fossil energy use [in Gigajoules/yr]	250 (234)	287	5	56
Fresh water withdrawal [in m^3 /yr]	1,688	1,868	612	644
Ecological Footprint [ha./person]	**4.3**	**5.1**	**0.4**	**1.8**

other sectors, such as manufacturing and service industries, ecologically balanced trade is assumed: the embodied energy and resources in exports is assumed to be equal to that in imports. A more in-depth analysis, however, would be fully corrected for any ecological trade imbalance (Footprint of imports minus Footprint of exports), data permitting.

The second step in Ecological Footprint analysis involves converting consumption to a corresponding land area for each consumption category. This requires that we know the ecological productivity for each land-use category. We use trade and productivity figures compiled by the UN Food and Agriculture Organization (FAO) to determine global average productivity for croplands. The productivity and carrying capacity of pasturelands was estimated from agricultural handbooks. Average forest productivity was set at 2.3 cubic metres of usable wood fibre per hectare per year. This corresponds to the average productivity of temperate Canadian forests,[20] and is also close to the 2 cubic metres per hectare per year used by the Dutch *Friends of the Earth* for analyzing global carrying capacity constraints.[21] As discussed above, we account for CO_2 sequestration from the burning of fossil fuel at a land-for-energy ratio of one hectare per 100 gigajoules. (To date we do not include the absorption land requirements of other forms of waste and pollution. Our EF calculations therefore underestimate the actual land demand of the consumption cycle.) Sample estimates of land "consumption" by Canadians are provided in Box 3.3.

As noted, the figures in Table 3.3 show the land areas required to provide the current lifestyle of an average Canadian. Thus, if we read across row *"43-books/magazines"* to the *"F-FOREST"* column, we find that 0.1 hectare of forest land are required to produce his/her reading materials. In addition, the embodied energy land associated with books and magazines is 0.06 hectare. This means that on average 0.16 hectare of land is required continuously to produce the fibre for each Canadian's newsprint consumption. The bottom

**Figure 3.4: The Footprint of the Average Canadian spreads over many
land categories and measures over four hectares.**

right corner of the matrix shows the total land required by average Canadians
— the *per capita* Ecological Footprint — to be 4.27 hectares, 2.34 hectares for
carbon dioxide assimilation alone. As shown in Table 3.5, the U.S. individual
footprint is proportionately larger at 5.1 hectares.

2) *How large is the Vancouver regional Footprint?*

With these data on the *per capita* Footprint, we can estimate how much global
hinterland the human inhabitants of typical industrialized regions appropriate
to maintain their much-vaunted material standards. We will use the Lower
Fraser Valley, which extends eastward 144 km from Vancouver, B.C., as our first
illustration. This urban-agricultural region, located between the U.S. border to
the south and mountains to the north, extends over approximately 4,000 square
kilometres (400,000 hectares) of habitable valley bottom and is home to
1,800,000 people, for a population density of 4.3 people per hectare. Assuming
average Canadian consumption patterns, our estimates of corresponding land
requirements show that the regional population has an actual Ecological
Footprint of 77,000 square kilometres (7,700,000 hectares). In other words, the
Lower Fraser Valley population requires an area 19 times larger than its home
territory to support its present consumer lifestyles, including 23,000 square
kilometres for food production, 11,000 square kilometres for forestry products,

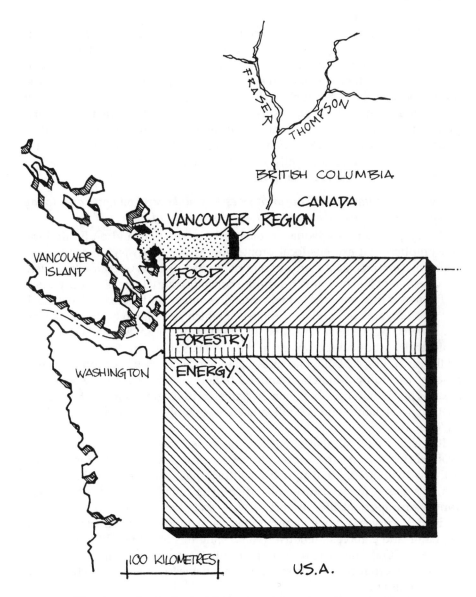

Figure 3.5: The Ecological Footprint of the Lower Fraser Valley.
The residents of British Columbia's most populated and ecologically most
productive region, the Lower Fraser Valley (stippled area), "appropriate" through
trade and natural ecological flows the productivity of an area 19 times the size of
their home region (hatched area) to satisfy present consumption levels of food,
forest, products and fossil fuel.

and 42,000 square kilometres to accommodate energy use (Figure 3.5). This figure represents the region's "ecological deficit" with the rest of the world.

Even if the land in the Lower Fraser Valley were twice as productive as global averages, the people of this region would still require the ecological output of nine times as much such land as is locally available. In fact, our colleague Yoshihiko Wada's analysis of regional output shows that agricultural productivity in the Valley is among the highest in Canada and more than double the world's average. However, average forest productivity over a Lower Mainland region slightly larger than that described above is about equivalent to world average output.

3) A global comparison of Footprint sizes — could everybody on Earth today enjoy North Americans' current ecological standard of living?

As long as there is adequate ecologically productive land on Earth, local consumption that exceeds local production in any given region can be sustained by "importing" the surplus output of other regions. Of course this raises the question of just how much surplus ecoproductivity there is on the planet.

To answer this, we need to know first how much land there is. Planet Earth has a surface area of 51 billion hectares of which 13.1 billion hectares are land not covered by ice or fresh water. Of this just under 8.9 billion hectares are ecologically productive: cropland, permanent pastures, forests and woodland. Of the remaining 4.2 billion hectares, 1.5 billion hectares are occupied by large deserts (excluding Antarctica) and 1.2 billion hectares by mostly semi-arid areas. The remaining 1.5 billion hectares include grasslands not used for pasture, wastelands, and 0.2 billion hectares of built-up areas and roads (0.03 ha/capita).

It seems, therefore, that about 8.9 billion hectares of land are potentially available for human exploitation. However, approximately 1.5 billion hectares of this is "wilderness" that arguably should remain in its near-pristine state. In fact, this mostly forested area is already fully engaged as it stands providing a variety of important life-support services and should not be otherwise exploited. It serves, among other things, as a biodiversity reserve, climate regulator, and carbon storehouse. (Harvesting this forest would lead to a net release of CO_2.) This would mean that only 7.4 of the 8.9 billion hectares of ecologically productive land are actually available for other more active forms of human use.[22]

Since the beginning of this century, the "available" *per capita* ecological space on Earth has decreased from between 5 and 6 hectares to only 1.5 hectares. Meanwhile, as material welfare has increased, the Ecological Footprints of people in some industrialized countries have *expanded* to more than four hectares (see Figure 1.5). These opposing trends illustrate the fundamental conflict confronting humanity and the real challenge of sustainability today:

the Ecological Footprints of average citizens in rich countries exceed their "fair earthshares" by a factor of two to three; thus, if everybody on Earth enjoyed the same ecological standards as North Americans, we would require three Earths to satisfy aggregate material demand using prevailing technology. While this may seem like an astonishing result, our underlying assumptions

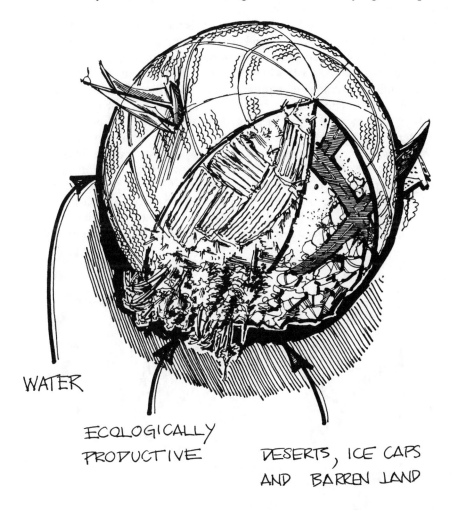

WATER

ECOLOGICALLY
PRODUCTIVE

DESERTS, ICE CAPS
AND BARREN LAND

**Figure 3.6: How Much Land Can Perform Functions Vital to Human Activities?
The Earth has a surface area of 51 billion hectares, of which 14.5 billion are land.
However, only 8.9 billion hectares of the land area are ecologically productive.
The remaining 5.6 billion hectares are marginally-, or un-productive, for
human use, 1.4 billion of which are covered by ice.**

and empirical evidence suggest that our calculated Footprint areas are actually considerable underestimates. In short, there are real biophysical constraints on material growth after all. Not even the present world population of 5.8 billion people — let alone the 10 billion expected by 2040 — can hope to achieve North America's material standard of living without destroying the ecosphere and precipitating their own collapse.

These findings beg another question: what is the present aggregate demand by people on the ecosphere? A rough assessment based on four major human requirements shows that current appropriations of natural resources and services already exceed Earth's long-term carrying capacity. Agriculture occupies 1.5 billion hectares of cropland and 3.3 billion hectares of pasture. Sustainable production of the current roundwood harvest (including firewood) would require a productive forest area of 1.7 billion hectares. To sequester the excess CO_2 released by fossil fuel combustion, an additional 3.0 billion hectares of carbon sink lands would have to be set aside. This adds up to a requirement of 9.6 billion hectares compared to the 7.4 billion hectares of ecologically productive land actually available for such purposes. In other words, these four functions alone exceed available carrying capacity by close to 30 percent. (Even if all 8.9 billion hectares of ecologically active land were included,[23] present "overshoot" exceeds 10 percent.) These simple footprint

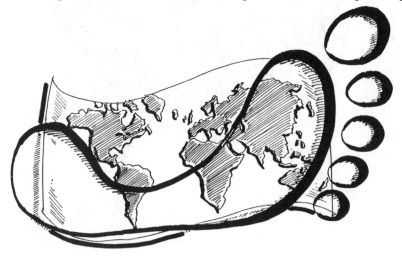

Figure 3.7: Humanity's Ecological Footprint is as much as 30 percent larger than nature can sustain in the long run. In other words, present consumption exceeds natural income by 30 percent and is therefore partially dependent on capital (wealth) depletion. The lavish partying by the wealthy today means a hefty bill for everyone tomorrow.

extrapolations alone are enough to suggest that present levels of material throughput in the global economy cannot be sustained. The global ecological deficit, unlike those of individual regions, obviously cannot be financed by trade; it depends instead on the liquidation of natural capital stocks. (Many people have concluded as much intuitively from reading the newspapers in recent years.)

What can we make of present international development objectives in light of these findings? The major agreed goal is to raise the developing world to present First World material standards. The Brundtland Commission, for example, argued for "more rapid economic growth in both industrial and developing countries" and suggested that "a five- to ten-fold increase in world industrial output can be anticipated by the time world population stabilizes some time in the next century."

Let's examine this prospect using Ecological Footprint analysis. If the present world population requires at least 9.6 billion hectares to sustain its activities, a five- to ten-fold increase would correspond to a total productive land requirement of 48 to 96 billion hectares (assuming the use of present technology). Thus, to accommodate *sustainably* the anticipated increase in population and economic output of the next four decades we would need six to twelve additional planets. The only alternative, if we continue to insist on economic growth as our major instrument of social policy, is to develop technologies that can provide the same levels of service with six to twelve times less energy and material. This is indeed a daunting task considering that the energy consumption of average households in industrialized countries is still increasing. One thing is certain, however: we cannot sustain development on phantom planets!

4) *Footprinting Great Britain*

"What is 120 times the size of London? The answer: the land area or *ecological footprint* required to supply London's environmental needs." So begins the *Executive Summary* of a detailed study completed early in 1995 by the International Institute for Environment and Development (IIED) for the U.K. Department of Environment and Development.[24] This report summarizes the use of the ecological footprint concept in Britain to date and explores its potential contribution to official policy development and action by non-government organizations. Specific objectives were to develop a U.K. context for the footprints debate "through case studies of some of the ecological footprints generated by U.K. consumption, production, trade and investment patterns in foreign coutnries; and "to explore how this information could be communicated to the public so that citizens can take positive actions which both lighten and reduce Britain's footprints *and* promote more sustainable development in poor and vulnerable developing countries" (page 7, original emphasis).

The IIED report notes that in the last 200 years concern about the distant impacts of British lifestyles has never been far below the surface. As the leading imperial power during the industrial revolution, "Britain was able to draw on resources from every corner of the globe to a historically unprecedented degree." Thus, in the early 19th century, the fictional traveler in Robert Southey's *Letters from England* noted that "all parts of the world are ransacked for the Englishman's table;" similarly, in modern times, George Orwell wrote in his depression era classic *The Road to Wigan Pier* that "in order that England may live in comparative comfort, a hundred million Indians must live on the verge of starvation — an evil state of affairs, but you acquiesce in it every time you step in a taxi or eat a plate of strawberries and cream" (page 9).

While the colonial era has long since passed, the IIED analysis shows that the "evil state of affairs" characterizing the ecological relationships among countries of differing economic status is as entrenched as ever. Several case studies are developed to reveal the extent of Britain's dependence on presumptive surplus capacity in other countries. For example, with its 10 kg *per capita* annual consumption of bananas, the U.K.'s population of 58 million consumes 580,000 tons of the tropical fruit per year. Assuming an average yield of 12 tonnes/hectare, then the "base" U.K. banana footprint is about 48,300 hectares (page 49). This is equivalent to two percent of the forest and woodland in Britain. Similarly, British consumption of 1,160,000 tonnes of raw cotton annually is estimated to appropriate 892,300 hectares of "high average yield cropland (1300 kg/ha) or 1,450,000 hectares of average African cotton-growing land (800 kg/ha)." This latter figure is equivalent to 22 percent of the U.K.'s domestic cropland (page 64).

The ecological footprint of the U.K.'s forest product imports is perhaps the most telling case examined. This analysis compiled the volume of different categories of wood-product imports, converted these to roundwood equivalents, and estimated the proportion originating in different countries and from different classes of woodland to estimate Britain's base impact on the world's forest. The results show that 6.4 million hectares of forest land throughout the world are taken up "more-or-less permanently" and that an additional 67,000 ha are deforested each year to provide wood products for the U.K. (75 percent of the latter in developing countries). Britain's total forest products footprint is three times the area of the country's own productive forest — in other words, only a quarter of the productive forest upon which the country depends is actually in the U.K.!

Using sustainable yield criteria, IIED estimates the U.K. *per capita* consumption of timber products is 66 percent higher than the "permissable" global average (i.e., the ecological fair share) (pages 82-83).

As impressive as these data are, IIED emphasizes (as we do for our analyses) that the base case results are likely to be considerable underestimates of the true footprints of British consumption. For example, "if one were to include a forest area required to provide the U.K. with other goods and services in addition to wood products — carbon sequestration to make up for the U.K.'s own burning of fossil fuels, carbon storage, essential biodiversity security, etc. — then it would be clear that the U.K. depends upon a foreign forest area even greater than three times its own forest area." If we added in the forests degraded by the U.K.'s pollution, then the footprint would be larger still. "For example, 20 percent of Norway's forests are suffering pollution-related die-back, and many of the pollutants originate in the U.K." (page 85).

Indeed, a major strength of the IIED study is its emphasis that base footprint calculations (the land areas appropriated for production of raw materials) do not account for myriad related direct and indirect effects, "the wider environmental and social impacts, such as waste, pollution and damage to health which are the most interesting aspects of the impacts of [banana production] on the environment" (page 50). This is seen as a significant contribution to the footprint debate: "It is important to examine the depletion of both human and environmental capacities for sustainable development: the 'ecological' focus of the footprint term obscures the fact that Britain's impact overseas can be as much to disrupt livelihoods as to degrade natural ecosystems" (page 33). Accordingly, in the U.K., "the footprint idea is now being used as an umbrella term to describe the whole range of impacts generated by one country's policy, consumption, production and investment patterns on the capacity of other countries to achieve sustainable development" (page 30).

We heartily concur with this general thrust. Our goal in advancing the ecological footprint concept is to critique the prevailing development paradigm and to help extend the international development debate beyond its focus on GDP growth to ecological reality. Now that real headway is being made in that direction, it is gratifying that users are emboldened to argue beyond ecology for ever more wholistic approaches to sustainability.

5) European examples: The Ecological Footprints of the Netherlands and the Trier Region of Germany

Let's compare Canada's Lower Fraser Valley (Example 2) with some European regions. With an area of 34,000 square kilometres and a population of 15 million, the population density of the Netherlands is 4.4 people per hectare. This is about the same density as in the Lower Fraser Valley. While the average Netherlander consumes fewer resources than the average Canadian, Holland still uses over 15 times more land than lies within the country's own political boundaries: the Dutch use approximately 5,400 square kilometres of built-up area, 100,000 square kilometres for food production for domestic consumption, 70,000 square kilometres for forestry products, and would need 320,000 square kilometres for CO_2 absorption (see Box 3.4). Indeed, Dutch government data

Figure 3.8: For urbanization, food, forest products and fossil fuel use, the Dutch use the ecological functions of a land area over 15 times larger than their country.

suggest that for fodder alone (including that used to produce food products for export) the Netherlands appropriates 100,000 to 140,000 square kilometres of arable land, much from the Third World. This is five to seven times the area of agricultural land in the entire country.[25] These estimates suggest that the ecological deficit of the Dutch is somewhat smaller but of the same order of

---

BOX 3.4: Assessing the Footprint of the Netherlands[26]

To keep things simple, we consider only four important categories of domestic consumption: built-up land, food, forest products and fossil energy. This avoids any significant double counting, yet is sufficient to illustrate the strength of Ecological Footprint analysis.

BASIC DATA **(Netherlands):**
  1991 population: 15,050,000; land area: 33,920 square kilometres.
  Built-up land: 538,000 hectares.
  Commercial energy consumption in 1991: 3,197 PJ – 36 PJ from non-fossil fuel sources (mainly nuclear energy). Therefore, for this calculation, (3197 – 36)[PJ] / 15 million Dutch =) 210 GJ/cap./yr. is used to represent the fossil fuel consumption.

CALCULATIONS:
  forest: assuming a consumption of 1.1 m^3/cap./yr. and a forest productivity of 2.3 m^3/ha/yr., this consumption corresponds to (1.1 [m^3/cap./yr.] / 2.3 [m^3/ha/yr.]) = 0.47 [ha/cap.] of forest land.
  fossil fuel: 210 [GJ/cap./yr.] corresponds to (210 [GJ/cap./yr.] / 100 [GJ/ha/yr.] =) 2.10 [ha/cap.].

RESULTS:

food:	cropland	0.45 [ha/cap.]
rangeland:		0.26 [ha/cap.]
forest:	1.1 [m^3/cap./yr.] corresponds to	0.47 [ha/cap.]
fossil fuel:	210 [GJ/cap./yr.] corresponds to	2.10 [ha/cap.]
degraded land (settlements and roads):		
	(538,000 [ha] / 15,000,000 [Dutch people])	0.04 [ha/cap.]
Total Individual Footprint:		**3.32 [ha/cap.]**

The Netherlands' aggregate Ecological Footprint is:
(15,000,000 [Dutch people] x 3.32 [ha/cap.] x 0.01 [ha/km²] =) 498,000 square kilometres. This is almost 15 times larger than the Dutch territory of 33,920 square kilometres.

---

magnitude as that of residents of the Lower Fraser Valley.

Even relatively rural regions of wealthy European countries run significant ecological deficits. The Trier region of Germany has a population density of only one person per hectare. Nevertheless, geography student Ingo Neumann of Trier University has estimated a *per capita* land requirement of over three hectares and a regional Ecological Footprint three times larger than the local land base. In short, large ecological deficits are the rule for industrialized regions and countries. The material prosperity of most industrialized regions depends heavily on extra-territorial ecoproductivity.

6) A Regional Analysis from Australia

A team led by Professor Rod Simpson at Griffith University in Brisbane has been working on the Ecological Footprint for the South-East Queensland (SEQ) region of Australia.[27] SEQ contains six of the fastest growing local government areas in Australia with growth rates in excess of 4.5 percent per year. The 1991 population was 1.85 million, expected to reach about three million by 2010. The total area of SEQ is 2.22 million hectares (5.50 million acres) of which 827,000 hectares (2.04 million acres) is under agricultural use.

Following the calculation procedures described in this book and using Australian, Queensland, and SEQ regional data as available, Simpson's team have estimated the *per capita* Ecological Footprint of the average resident of South East Queensland at 3.74 hectares. The use of Australian productivity data likely produces a smaller EF estimate than would be the case if the calculations were based on world averages as in our Table 3.3. Even so, some significant differences between average Canadian and SEQ consumption patterns emerge. For example, Australians — at least in SEQ — apparently use twice as much fossil energy in the transportation of goods as do Canadians, but consume considerably less wood in house construction and less operational energy in heating their homes. These contrasts are as expected from climatic and geographic differences between the two case studies and show the potential of EF analysis "in identifying important differences between regions."

Simpson's data show that the total EF for SEQ is about 6.91 million hectares (17.1 million acres). Thus the region's human population overshoots its local carrying capacity by a factor of 3:1. However, if we extrapolate Simpson *et al.*'s regional data to the entire country, Australia appears to have what may be the greatest surplus carrying capacity of any wealthy advanced country (see Box 3.4). As is the case for Canada, however, much of this seeming surplus is undoubtedly incorporated into the Footprints of other countries.

7) What does ecological dependence mean for trade?

People in Canada's Lower Fraser Valley make an Ecological Footprint 19 times the size of their home region; the Dutch "consume" a land area 15 times larger than their country; with an estimated 2.5 hectare Footprint per average Japanese, the Footprint of Japan would be more than eight times the size of that island country's productive land base. It seems that the Footprint of many industrialized regions and countries is about an order of magnitude larger than their political territories (Box 3.5). The quasi-parasitic relationship between

BOX 3.5: The Ecological Deficits of Industrialized Countries[28]

COUNTRY	ecologically productive land (in hectares) a	population (1995) b	ecol. productive land *per capita* (in hectares) c = a/b	NATIONAL ECOLOGICAL DEFICIT *per capita* (in hectares) d = Footpr. – c	(in % available) e = d/c
countries with 2–3 ha Footprints				*assuming a 2 hectare Footprint*	
Japan	30,417,000	125,000,000	0.24	1.76	730%
S. Korea	8,716,000	45,000,000	0.19	1.81	950%
countries with 3–4 ha Footprints				*assuming a 3 hectare Footprint*	
Austria	6,740,000	7,900,000	0.85	2.15	250%
Belgium	1,987,000	10,000,000	0.20	2.80	1,400%
Britain	20,360,000	58,000,000	0.35	2.65	760%
Denmark	3,270,000	5,200,000	0.62	2.38	380%
France	45,385,000	57,800,000	0.78	2.22	280%
Germany	27,734,000	81,300,000	0.34	2.66	780%
Netherlands	2,300,000	15,500,000	0.15	2.85	1,900%
Switzerland	3,073,000	7,000,000	0.44	2.56	580%
				assuming a 3.74 hectare Footprint	
Australia	575,993,000	17,900,000	32.18	(28.44)	(760%)
countries with 4–5 ha Footprints				*assuming a 4.3 (Canada) or 5.1 ha. (U.S.) Footprint*	
Canada	434,477,000	28,500,000	15.24	(10.94)	(250%)
United States	725,643,000	258,000,000	2.81	2.29	80%

This table shows that most industrialized countries run a significant ecological deficit. The last two columns represent low estimates of these *per capita* deficits. Even if their land were twice as productive as the world average, European countries would still run a deficit more than three times larger than domestic natural income. Canada and Australia are among the few developed countries that consume less than their natural income domestically. However, their natural capital stocks are being depleted by exports of energy, forest and agricultural products, etc. In short, the apparent essergy surpluses in these countries are being incorporated by trade into Ecological Footprints of other "advanced" countries.

these advanced economies and the rest of the world thus revealed by ecological footprint analysis is a predictable consequence of the entropy law (see Box 2.4). All such energy- and material-intense economies depend for their internal integrity on "essergy" imported from elsewhere.

Significantly, these and similarly industrially advanced regions are regarded as economic success stories. Each boasts a positive trade and current account balance measured in monetary terms and their populations are among the most prosperous on Earth. By contrast, our ecological analysis of physical flows shows these same areas to be running massive unaccounted ecological deficits with the rest of the planet. This raises difficult developmental questions in a world that holds such countries up as models for emulation and whose principal strategy for sustainability is economic growth. Global sustainability cannot be ecological deficit financed; not all countries or regions can be net importers of carrying capacity! This fact has profound implications for conventional development models.

Ecological Footprint assessments should enable policy decision-makers to understand better the long-term constraints facing national and international economies as populations and *per capita* consumption increase. For example, Ecological Footprint analysis can estimate the balance of trade in load-bearing capacity as embodied in the energy and material flows associated with trade goods and biogeochemical cycles. This can reveal costs and benefits of trade, including potential long-term sources of interregional conflict, to which con-

---

BOX 3.6: Calculating the Footprint of India[29]

DATA:

In 1994, India had a population of 910,000,000 people and a land area of 297,319,000 hectares. About 250,000,000 hectares are productive.

Indians consumed 4,900 PJ of fossil fuel in 1991 and 2,824 PJ of traditional fuels (mainly wood). Hence, the fossil fuel consumption is about 5 GJ/cap./yr., which corresponds to about 0.05 ha/cap.

Food production was close to self-sufficiency (at least in economic terms). There are 180,000,000 hectares of crop and pasture land. Imports add another 1,000,000 hectares (calculated by dividing the traded products by their respective ecological productivity: -344,000 [metric tonnes of cereals] / 2.5 [t/ha/yr.] + 376,000 [t of oil] / 1.0 [t/ha/yr.] + 583,000 [t of legumes] / 0.8 [t/ha/yr.]). Hence the *per capita* land area for food is 0.2 hectares.

Forestry: In 1991, roundwood consumption in India was about 281,045,000 m^3, or about 0.31 m^3/cap./yr. Assuming a sustainable forest productivity of 2.3 m^3/ha/yr., this amounts to a *per capita* land area requirement of 0.13 hectares, or a total of 118 million hectares. India has only 66.7 million hectares of forest and woodland..

RESULT:

food:	0.20 [ha/cap.]
forest:	0.13 [ha/cap.]
fossil fuel:	0.05 [ha/cap.]
Footprint:	**0.38 [ha/cap.]**

ventional monetary analysis is blind. Comparing a region's own productive capacity with its actual carrying capacity demand reveals a "sustainability gap" that is presently being bridged by potentially unsustainable imports or local natural capital depletion. Understanding this reality raises the question of the relationship between ecological security and geopolitical stability and should force a reconsideration of the proper role of trade.

The ecological deficits of "wealthy" countries may become a growing concern for those participants in the global economy (typically low-income countries with large resource sectors) whose surpluses are being appropriated. The ecological "leakages" of these countries are actually being encouraged by current terms of trade. However, ecologically imbalanced trade may also become an issue to those "advanced" economies that have become dependent on others' carrying capacity. How secure are ecological surpluses upon which developed regions currently depend? What does this dependency mean for the

India's estimated *per capita* footprint is 0.38 hectares. However, according to consumption distribution calculated by the Indira Gandhi Institute, the Footprint of the average person in the bottom 50 percent of income-earners would be about half the above national average or roughly 0.2 hectares per person.

India's national Ecological Footprint is (910,000,000 [Indians] x 0.38 [ha/cap.]) or approximately 346,000,000 hectares as compared to the available 250,000,000 hectares of productive land. Where do the missing 96,000,000 hectares come from? Some of it is "imported": 1,000,000 hectares in the form of food, and about 45,000,000 hectares in CO_2 assimilation capacity. The remaining 50,000,000 hectares corresponds to other imports and the depletion of domestic natural capital assets, particularly forest cover. In other words, if India requires the sustainable output of 346,000,000 hectares of land (= required natural income), then 96,000,00 hectares estimates the scale of India's domestic and international ecological deficit.[30]

We should emphasize at this point that small Ecological Footprints do not necessarily imply a low quality of life. In fact, Kerala, a southern state in India, has a *per capita* income of about one dollar a day (less than a sixtieth of North American incomes). However, life expectancy, infant mortality, and literacy rates are similar to those of industrialized countries. The people in Kerala enjoy good health care and educational systems, a vibrant democracy, and a stable population size. It seems that Kerala's exceptional standard of living is based more on accumulated social capital than on manufactured capital. The world has much to learn from the people of Kerala.[31]

Key: PJ = petajoules   GJ = gigajoules   /cap. = *per capita*
       t = metric tonnes (1000 Kg.)   yr. = year   ha = hectares
       pulses = edible seeds of legumes (e.g. beans, lentils)

potential intensification of local and global resource conflicts? What sorts of new international agreements are needed to formalize stable relationships among interdependent regions? As important, how can we protect common-pool life-support functions upon which we are all dependent from rising global demand when any country can freely appropriate them to excess? In a shrinking world, interregional dependency is potentially a stabilizing force; in present circumstances it is more likely to be *de*stabilizing.[32]

Such considerations challenge conventional economic development models as promoted by the World Bank, the International Monetary Fund (IMF), the World Trade Organization (WTO, formerly GATT), and the Harvard Institute for International Development on grounds that there is simply not enough biophysical capital to sustain prevailing development myths. Expressing the mainstream view, Michael Roemer from the Harvard Institute writes that "economic growth is the only mechanism through which the welfare of the poor can be improved in a sustainable way" (*The Economist*, 4 June 1994, p. 6). However, putting aside for the moment the assumption that welfare equates to GDP growth, this perspective ignores the fact that for sustainability in a "full" world, consumption by the rich must be reduced to create the ecological space for increased consumption by the poor. Clearly, unqualified expansionist models that encourage expanded trade (and improved access to the world's resources by the industrialized countries) both promote dangerous illusions of universal prosperity and obscure the *de facto* direct competition between the rich and the poor for declining global carrying capacity.

8) Is a person's Ecological Footprint related to income?

Conventional economic development wisdom assumes there are no serious constraints on economic expansion, and that poverty can be alleviated most easily by increasing economic production. This perspective is attractive because it implies that people already enjoying high consumption levels do not have to compromise their lifestyles so that those in need can improve their material standards. In fact, many analysts even argue that more consumption by the rich benefits the poor since it accelerates growth and creates jobs by expanding the export markets of developing countries. As implied in a metaphor used by World Bank vice-president and chief economist Lawrence Summers, this view assumes "...rising tides do raise all boats."[33]

Limits to growth are invisible to static monetary analyses because monetary expansion itself is not bound by physical limits. It seems as if everyone can have all s/he wants provided s/he can pay for it. The ecological perspective, however, challenges this money-based view. Clearly the physical consumption of natural income by one person pre-empts any other person from using those same income flows. If global carrying capacity has been exceeded, then consumption by the rich *already* undermines prospects for the poor. In these

Figure 3.9: Equity. In today's ecologically overloaded world,
we are all in competition for the finite flow of natural income
produced by the ecosphere. In Ecological Footprint terms,
excess consumption by affluent countries occupies ecological space that would
otherwise be available to the poorer nations. Even within countries individual
Footprint sizes vary significantly.

circumstances, further GDP growth using existing technologies would necessarily require unsustainable depletion of natural capital and the further overfilling of waste sinks.

Ecological Footprint analysis reveals the growing competing demands on natural capital and thus raises the issues both of equity and the long-term sustainability of production. In the current global economy with its increasingly uniform international monetary system, those with the greatest financial clout gain the greatest and fastest access to the limited resource stocks of the world. The resultant economic growth does lead to further accumulation of human-made wealth, but in relatively few hands, and little is reinvested in maintaining the natural capital base on which so many money fortunes are founded. In short, the world's money rich simultaneously make the largest claim on natural income (i.e., they have the largest Footprints) even as their actions (inadvertently?) reduce future productive potential for all.

According to UN statistics, the 1.1 billion people who live in affluence consume over three-quarters of the world's total output. The remaining 4.7 billion people — 80 percent of the population — survive on less than a quarter of world output. However, as noted previously, preliminary estimates suggest that the Ecological Footprint of food, forest products and fossil fuel consumption alone already exceeds global carrying capacity by an unsustainable 30 percent — consumption by the affluent 1.1 billion people alone claims more than the entire carrying capacity of the planet. Ecological analysis thus highlights the ethical dimension of the sustainability dilemma and undermines reliance on sheer growth as a remedy for poverty. More particularly, it may force us to face the possibility that for sustainability the affluent will have to reduce their present share of consumption so the impoverished can increase theirs.[34] The question is: does the human family have the moral and political will to negotiate a global social contract governing more equitable access to ecological goods and services for all the world's people?

Ecological Footprint analysis can provide other useful perspectives on comparative consumption. For example, even within rich countries there is considerable income disparity, which is reflected in consumption patterns. Preliminary estimates suggest that in Canada the poorest 20 percent of the population have an average Ecological Footprint of less than three hectares while the richest 20 percent consume the ecological goods and services of over 12 hectares *per capita*.

While the wealthy have larger Footprints than the less well off, they also have more lifestyle choices that affect the size of their Ecological Footprints. For example, they can afford either large houses on the suburban fringe, which make it necessary to commute long distances, or they can live in expensive urban town-houses near their work, which reduces material, transportation and associated energy costs. Most people, however, do have some flexibility in their consumption patterns, which can reduce their Ecological Footprints. Locally produced food, organically grown vegetables, improved insulation, using bicycles and public transit, etc., all have smaller Ecological Footprints per dollar spent than the usual alternatives.

9) Housing choice affects our Footprints

A person's Footprint size is not fixed by income — it also depends on spending patterns. In many cases, housing type and location are the chief determinants as they influence house size and the household's transportation requirements. Living in densely populated urban areas leads to smaller *per capita* Footprints because of more efficient land-use and infrastructure and reduced transportation and residential heating requirements. A recent study of the San Francisco region found that doubling residential density cuts private transportation by 20 to 30 percent; urbanist Peter Newman reports differences in heating consumption between grouped and free-standing housing of up to 50 percent.[35]

Figure 3.10: If we choose to live at higher densities we can shrink our Ecological Footprints. In the process, we may find that our cities can become more community-oriented and livable places.

Figure 3.11: Mirrored Density: To assess a household's contribution of density to a city's resource consumption, we extrapolate the energy and material requirements of sample housing types to the entire city as if all households lived that way. This clearly contrasts the Footprints of different densities and lifestyles at the whole city level. (Using average figures for whole neighborhoods dilutes the effects of different lifestyle choices.)

Lyle Walker has examined the implications of housing and other lifestyle choices on the Ecological Footprints of households in various income groups for the UBC Task Force on Healthy and Sustainable Communities. He estimated the effects of income, housing type/density, and transportation options for households in the Vancouver region. The link between different housing types/densities and corresponding transportation requirements was established by extrapolating baseline data for particular housing types to the entire city (the "mirrored density" approach) and then estimating the transportation needs of such a hypothetical city (Figure 3.11).

Preliminary estimates show that living in a multi-unit condominium or apartment of similar market value to a suburban house, and using a compact, energy-efficient car rather than a standard-sized vehicle, can reduce a household's transportation and housing Footprint by a factor of three. At the same time, the condominium dwellers may see an improvement in their quality of life — they may be able to walk to work, be closer to friends and relatives (more contacts per square kilometre), enjoy a more vital neighborhood, and take advantage of greater recreational diversity such as parks, pedestrian areas, street cafés, movie theatres, and more.

Our new urbanites might also be able to achieve a threefold Footprint reduction in the food and consumer goods categories without significantly compromising quality of life. This would require reducing the animal-based component of the typical North American diet and relying on less processed and packaged food (both of which measures may produce health benefits). When purchasing goods, they would have to place more emphasis on quality and durability. In this sense, we should all become a little *more* materialistic — we should care that our utensils and toys last so we don't have to discard them prematurely.

An obvious application of this procedure might be to compare the Ecological Footprint statistics of different municipalities. Keep in mind, however, that comparisons of municipal-Footprint to municipal-land-area ratios can be misleading. The more densely populated municipalities will show proportionately larger ecological deficits while, in fact, requiring smaller *per capita* Footprints. It might be more informative, therefore, to document the average *per capita* Footprints of subject municipalities for comparisons among municipalities and to national and international averages. Other interesting comparisons include that of a defined population's Footprint vs. the actual area of its home region (see Lower Fraser and Dutch examples above), or among the estimated Footprints of alternative municipal development plans to assess the effects of different population densities or transportation technologies. (A municipality can substantially influence the size of people's Footprints by making high density more vital and attractive, and planning a less car-dependent transportation system — see Chapter 4.)

Industrialized countries' Ecological Footprints are particularly large due to these countries' voracious use of fossil fuel. Any engineer will testify how inefficiently we consume this valuable but underpriced resource. Heating domestic water and houses with fossil fuel is inexpensive but produces a large

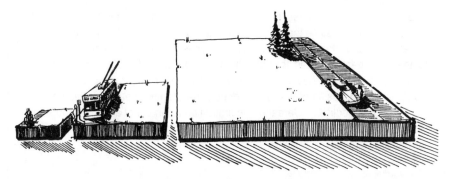

Figure 3.12: The Ecological Footprint of one person traveling five kilometres twice each workday (10 km per week) varies according to transportation mode: for bicycles, it is about 122 square metres, for buses 301 square metres and for cars 1,442 square metres.

contribution to the Footprints of households. One of the most promising strategies for decreasing our domestic Footprints, therefore, involves switching to renewable energy sources. For example, heating a given amount of water with solar-thermal collectors results in a Footprint 100 times smaller than heating with fossil energy (see above in this chapter).

10) How much ecologically productive land supports commuting by bicycle, bus and car?

BOX 3.7: Determining the Footprint of Commuting

REFERENCES AND ASSUMPTIONS:
The U.S. has 15 million hectares of roads, mainly for cars (and mostly built on agricultural land).
There are 1.75 people per car in the U.S.
We assume 230 workdays per year.
According to the Bicycle Federation of America, a bicycle rider requires 900 kJ of food per 10 kilometres.
According to Environment Canada, cars make up 98.4 percent of the traffic in Vancouver rush hour, while carrying only 62 percent of the commuters. Hence, we can conclude that a bus passenger needs only about 2.6 percent of the road space that a car driver occupies.
(For mathematical wizards, the calculation is as follows: (0.016/.38) / (0.984/.62) = 0.026)

CALCULATIONS:
Bicycle: The bicycle rider requires an extra 900 kJ/day of food for this daily 10 kilometre trip. We assume that this extra energy stems from breakfast cereals. These cereals need land to grow and energy for processing. The land equivalent of the commercial energy needed for agricultural production and for food processing of plant crops is typically the same as the crop area; hence, the total land area for the growing and processing of the food is double the growing area. The road space is assumed to be negligible. Cereals have a nutritional content of about 13,000 kJ per kilogram. The world average in agricultural production is 2,600 kilograms of cereals per hectare per year.

$$\frac{900 \text{ [kJ/cap./day]} \times 230 \text{ [days/yr.]} \times 2}{13,000 \text{ [kJ/kg]} \times 2,600 \text{ [kg/ha/yr.]}} = 0.0122 \text{ hectares or } \mathbf{122 \text{ square metres}} \text{ per rider}$$

Let's look at relative commuter virtue by using Footprint analysis to compare the ecological efficiency of cars, buses and bicycles. It turns out that a person living five kilometres from work requires an extra 122 square metres of ecologically productive land for bicycling, 301 square metres for busing, or 1,442 square metres for driving alone by car. The land for the cyclist is needed to grow extra food while most of the bus passenger's and car driver's land is taken up by CO_2 sequestration (Box 3.7).

Car: The average direct gas consumption by American cars is about 12 litres per 100 kilometres; indirect carbon consumption for car manufacturing and road maintenance adds 45 percent. Each litre of gasoline contains about 35 megajoules or 0.035 gigajoules of energy. Therefore, the fossil fuel Footprint of auto commuting is:

$$\frac{1.45 \times 12 \, [l/100 \text{ km}] \times 0.035 \, [GJ/l] \times 10 \, [\text{km/day}] \times 230 \, [\text{workdays/yr.}]}{100 \, [\text{km}] \times 100 \, [GJ/ha/yr.]} = 0.14 \, [\text{ha/cap.}] = 1,400 \, [m^2/\text{cap.}]$$

In addition, cars need road space. The road space per U.S. citizen is:

$$\frac{15,000,000 \, [\text{ha}]}{250,000,000 \, [\text{Americans}]} = 0.06 \, [\text{ha/cap.}] = 600 \, [m^2/\text{capita}]$$

Cars use 97.4 percent of the road space. However, the daily (2 x 5 [km] =) 10 [km] commute represents only about 1/8 of average annual car usage, and every car represents 1.75 people. Therefore, the *per capita* road space required for the 10 kilometre commute would be (0.974 x 1/8 x 600 / 1.75) = 42 [m²], mostly on agricultural land. Therefore, the total land appropriation for single occupancy car commuting sums to **1,442 square metres** of land.

Bus: The energy requirement of short-distance buses is 0.9 [MJ/cap./km]. Indirect energy requirements for roads, buses and maintenance are assumed to add an additional 45 percent (same as cars).

$$\frac{1.45 \times 0.0009 \, [GJ/\text{cap./km}] \times 230 \, [\text{workdays/yr.}] \times 10 \, [\text{km/d}]}{100 \, [GJ/ha/yr.]} = 0.03 \, [\text{ha/cap.}] = 300 \, [m^2/\text{cap.}]$$

In addition, buses need road space. As a first approximation (see above) we assume that a bus passenger would only use 2.6 percent of the road space a car driver requires for the same distance, i.e., (42 x 0.026 =) 1 [m²]. Therefore, the total land appropriation for the two daily 5 [km] bus rides requires approximately **301 square metres**.

Figure 3.13: The ecological efficiency of various technologies can be assessed by Ecological Footprinting. This example compares two ways of growing tomatoes in British Columbia: open-field production and heated hydroponic greenhouses. Even though a greenhouse's physical Footprint per unit production is much smaller than that for open-field production, when we consider the Ecological Footprints of energy, fertilizer and other inputs, the greenhouse's total land requirement per tomato is actually 10 to 20 times larger (drawn to scale).

11) Did you know that tomatoes leave Footprints?

Ecological Footprinting can be used to compare the "resource intensity" of competing technologies. In other words, we can use it to ask whether some new technology is really an ecological improvement over one it replaces. For example, Yoshihiko Wada compared the Ecological Footprints of two different high-tech approaches to producing a given quantity of tomatoes. Wada's master's thesis asked how much "embodied" land is needed to grow tomatoes in heated greenhouses compared to intensive field agriculture.[36] In each case, the total area would comprise the land directly occupied by the farming operation as well as the land equivalent of all the material and energy inputs (including space heating in the case of greenhouses) used to maintain production.

In terms of growing area alone, hydroponic greenhouses appear to be seven to nine times more productive than field cropping. However, Wada's work shows that heated hydroponic greenhouses in British Columbia actually require 10 to 20 times *more* Ecological Footprint per kilogram of tomatoes harvested than even high-input open-field production. It seems that hydroponic greenhouses are rather like factories, assembling tomatoes from a variety of energy intensive parts and inputs!

This case illustrates the difference between apparent economic and actual ecological efficiency: hydroponic greenhouses typically outperform open-field production economically but require much greater inputs. Thus, because of resource underpricing and unaccounted ecological costs, producing vegetables in greenhouses remains economically rational and may seem a viable alternative to preserving farmland. Full cost accounting would present a rather different picture. This example illustrates that economic success can be misleading and is certainly not always compatible with ecological integrity.

12) *The Ecological Footprint of bridges*

Ecological Footprinting adds a new dimension to the "environmental impact assessment" of major capital projects. All large-scale developments such as power plants and transportation infrastructure projects and even changes in zoning can have long-ranging effects on material and energy consumption, which are usually ignored in traditional environmental impact assessment. Ecological Footprint analysis can illustrate not only the direct impacts of resources used for such projects, but also the indirect effects resulting from the

Figure 3.14: The Ecological Footprints of Bridges.
New or expanded bridges for private cars stimulate urban sprawl and further car dependence, expanding the Footprint of the user population.

lifestyle changes they facilitate.

Students at Simon Fraser University in Burnaby, B.C. have worked out two cases involving the indirect ecological impacts of proposed bridges. Both studies asked how much additional ecologically productive land would be appropriated to accommodate the changes in living and consumption patterns that would be induced by the bridge projects. Gavin Davidson and Christina Robb analyzed the implications of widening the Lions Gate Bridge between Vancouver proper and the "North Shore" from three to five lanes. Using conservative assumptions (a constant regional population, no increased recreational traffic, no impacts outside Vancouver and the North Shore municipalities of North and West Vancouver) they concluded that changes in the settlement and transportation patterns that would be induced by the additional two lanes would add 200 square kilometres of ecologically productive land to the Ecological Footprint of the study area.

The second study by David Maguire, Calvin Peters and Marcy Saprowich used economic projections from the Federal Environmental Assessment Review Office in Ottawa to estimate the likely Ecological Footprint of the infamous "fixed-link" between the mainland and the island province of Prince Edward Island in Eastern Canada. They concluded that constructing the bridge and replacing the existing ferry service would lead to the additional appropriation of approximately 160 square kilometres of ecologically productive land.

These examples show how Footprint analysis can contribute to the assessments of alternative technologies and capital projects by revealing heretofore hidden implications for resource consumption and waste generation. This in turn suggests that EF analysis might usefully be applied to less tangible areas such as policy, program and budget assessment. The relevant question in an ecologically full world is: to what extent will this policy (or program or budget) contribute to increasing or reducing the Ecological Footprint of the affected population? An early step in any such analysis would be to document all the possible direct and indirect effects of the policy on resource consumption and waste generation (using, for example, a modified form of input-output analysis). Particularly important would be policy-induced changes in the lifestyles of the affected population. Quantifying these effects, converting them to land equivalents, and summing them after correcting for double counting would reveal the incremental (or decremental) change in the Ecological Footprint of the subject populations that might be expected from the policy innovation.

Clearly the parameters for EF analysis will vary among applications. Each case involves different systems boundaries and hidden indirect effects, and some may have totally unique features. All these factors are subject to personal judgment and values. However, it is a strength of EF analysis of public policy that it forces the analysts both to explore critical issues and impacts that have been ignored to date and to declare their judgments and values in reflecting upon — and deciding on — the new trade-offs revealed in making their policy and development decisions.

13) Learning about sustainability in schools and in the outdoors

Ecological Footprinting is a stimulating way to introduce students to some of the less obvious but crucial dimensions of human ecology and to familiarize them with some of the ecological implications of the consumer society. Several individuals and organizations are developing teaching tools and handbooks

Figure 3.15: The Ecological Footprint concept has been integrated into various outdoor and indoor educational activities. It can be used in games and school projects to study energy and material flows in nature, experiment with lifestyles, and provide local concrete applications for the mathematics, biology and physics being taught simultaneously.

to assist students and teachers in undertaking their own simple Footprint analyses.

- How big is one hectare of forest? Measure one hectare of forest as a square and put little flags at each corner. How long does it take to walk around this hectare? Count the number of trees more than two inches in diameter at breast height in this hectare. In some cases you may have to estimate the total from a sample of, say, one-tenth of the area. How would you estimate the average age of these trees? Try to find out from local records what the annual productivity of forests in your area is. Use these data to estimate the total volume of wood standing in your hectare. Use available conversion ratios to determine its carbon and energy content? Estimate how far you could drive an average car with the equivalent energy. If a car travels 20,000 km per year and uses one litre of fuel per 10 km, what is the Ecological Footprint of operating that vehicle in your forest region? (i.e., how many hectares of growing forest would be required to absorb the CO_2 emitted by the car?) What other values are found in the forest apart from its timber production and carbon sink functions?

- How large is *your* Ecological Footprint? What is the total Ecological Footprint of all the students in your school? Can you draw this Footprint on a map of the neighborhood? How does it compare to the entire area occupied by your school including the school yard and playing fields? Assuming average consumption levels (Table 3.3), estimate the Footprint of your municipality and draw it to scale on a map of your home region. How much larger than the political area of your town is its Ecological Footprint? Find out where major consumption items — e.g., lumber, cars, TVs, fridges, clothes, food — available in your community come from and record their origins on a map of the world. In how many countries and continents have you found part of your community's Ecological Footprint?

These are just a few examples of how questions stimulated by the Ecological Footprint concept can kindle people's ecological curiosity and help them to find answers about human-nature relationships on their own. Such exercises also hone student's basic data gathering, research and quantitative skills. Properly designed exercises enable students to apply the EF concept and "experience" the quantities involved in real-world settings with which they are familiar (woodlots, school yard, home towns) rather than with abstract numbers or far-removed examples. Such exercises can sometimes provide a venue to integrate various other educational activities and subject areas such as biology, mathematics, physics, history, economic geography, and social studies.

Some specific examples: the Sea to Sky outdoor school in Gibsons, B.C., has integrated many Footprint applications into its programs. Participatory outdoor activities focus on the relationship between human consumption and ecological production, tracing food and goods to their origins, exploring how different human activities compete for nature's services, analyzing how socio-economic variables affect Ecological Footprint size, experimenting with low-Footprint lifestyle choices, and more; ESSA Technologies, an environmental consulting company in Vancouver, and the B.C. Ministry of Environment, Lands and Parks have prepared a teacher's guide to British Columbia's State of the Environment Report, which contains a section that takes students through an analysis of the food component of the Ecological Footprint; Jim Wiese with the educational involvement program of B.C. Hydro's "Power Smart" initiative has used Ecological Footprinting to illustrate energy use in daily life and to help students evaluate lifestyle choices and potential energy savings. He has developed a simple checklist on household energy consumption and consumer behavior to help students calculate the Ecological Footprint of their households.

14) State of the Environment Reporting

The Ecological Footprint can be used as an effective indicator of sustainability and ecological health. It therefore caught the interest of the Canadian *State of the Environment Report* team who were considering a shift in their conceptual approach away from an environmental indicators framework (as prominently used in the 1991 report) towards a more integrated human ecology perspective.

Figure 3.16: Ecological Footprinting is an ideal tool for sustainability reporting. It can compare countries or regions, analyze the ecological implications of trends and issues, or evaluate progress toward sustainability.

They commissioned Colin Duffield to develop ideas to incorporate applications of Ecological Footprint analysis into the 1996 report.[38] He suggested applying the concept to urban systems to compare the Footprints of urban, suburban and rural living; to analyze the ecological implications of trends towards smaller household sizes; and to compare alternative urban densities and transportation strategies. He also suggested desegregating the *per capita* Footprint to show the contributions of various human activities such as agriculture and heavy industry, and lifestyle choices such as diet and recreational activities. In a similar vein, the 1993 *Environmental SCAN* prepared for the Canadian Council of Ministers of the Environment and a report to the Fraser Basin Management Board in Vancouver, both prepared by accounting firm Peat Marwick Stevenson and Kellogg, proposed Ecological Footprinting as a way to "assess sustainability from an ecological world view."[39]

15) Interpreting sustainability: The ecological "Rorschach Test"...

Many community activists and other sustainability advocates overestimate the public's support for action toward a more sustainable society. Ambivalence or lack of support for action can partly be explained by general ignorance but is caused more often by psychological and institutional barriers, conflicts with other goals, entrenched belief in alternative models, or economic disincentives. This section discusses some of the strengths of Ecological Footprinting in overcoming some of the barriers to sustainability we have encountered in our research and in presentations to community groups. Understanding these barriers can help activists and planners tailor sustainability initiatives to specific conditions, thus increasing the probability of success.

We find that the Ecological Footprint concept helps us explore people's concerns and perceptions of sustainability. By clearly framing the ecological challenges, the Ecological Footprint concept facilitates constructive communication. Asking how we can reduce our Ecological Footprint while improving our quality of life builds a more productive basis for debate than conventional sustainability models.

Figure 3.17: Ecological Footprint analysis is a useful conversational tool in exploring people's concerns and challenging their assumptions about sustainability.

To illustrate, one popular mainstream model depicts sustainability as achieving a magic "balance" among the three intersecting circles of economy, society and environment. While this approach contributes to consensus-building, it may also lead to policies rooted in ambiguity and misunderstanding. By implying rough equivalency among the three spheres, the model avoids hard questions and contributes to social denial. For example, are the three spheres really equivalent and interchangeable? What *is* the functional relationship among them? Can we really substitute big chunks of nature with "an equivalent value" of human-made capital? Is the economy a means or an end in itself?

We believe the intersecting circles model obscures real imbalances, non-equivalencies, and moral issues critical to sustainability. For example, humanity needs the ecosphere, but the ecosphere does not need us. This is not to argue that economy and society are less important to humanity than ecology but rather that we need to understand the "directionality" of dependence before we can produce good policy for sustainability. Indeed, should we not be structuring the economy better to serve society rather than, as at present, restructuring society to serve the economy? A more accurate system of relationship can be depicted using concentric circles showing the humansphere — society and economy — as subsystem(s) of the ecosphere (see Figure 1.1). Once we acknowledge this hierarchical dependence, policy has a better chance to take us in the direction we all say we want to go. Rather than allowing the economy to drive society and displace the ecosphere as at present, the new model may force the focus onto the changes needed in society to recapture the economy and to govern both in harmony with the ecosphere. Most important, it recognizes that beyond certain points, there are no further fruitful "trade-offs" to be made with the ecosphere: expanding human infrastructure at the cost of increasing the load on nature impoverishes society today and places the future in jeopardy.

Starting from this perspective, the Ecological Footprint model advances a concrete and measurable bottom-line condition for sustainability: "humanity must live within the means of nature." It therefore facilitates more fruitful exchange among alternative perspectives on sustainability than conventional approaches. Recognition of real ecological constraints triggers debate about the meaning of sustainability, forces discussion of alternative approaches to "living within the means," illuminates barriers to action and ways to overcome them, and, most importantly, stimulates rethinking of the meaning of "quality of life."

Ecological Footprinting challenges common assumptions about economy, society, and nature and places the issue of over-consumption by affluent countries (and the rich everywhere) openly on the table. Nevertheless, it seems to resonate with many different audiences. People newly exposed to the Ecological Footprint model can easily discuss it in their own terms, and many are able to point out implications for the present economy. This quickly moves discussion beyond such usual sustainability concerns as local pollution, waste recycling, and conserving biodiversity somewhere else. Thus, in discussing Ecological Footprints, most people come quickly to reconsider human dependence on nature and the consequences of continued degradation of the ecosphere. Many accept that reducing throughput is a prerequisite for sustainability; acknowledge that various human uses of nature are in competition; and show willingness to reconsider the nature of economy-ecology linkages. They recognize that maintaining nature's capacity to regenerate is a necessary condition for sustainability and that this condition is not now being satisfied. It is a short leap from this to accepting the need for fundamental change. Certainly few continue to argue that without massive technological and institutional reform, everyone on Earth could live as North Americans do today.

However, few people support the notion that wealthy countries must reduce their consumption of resources. This may prove to be the major bottleneck on the road to sustainability: even though we acknowledge the constraints on an abstract general level, we are reluctant to gamble on the admittedly uncertain personal consequences of required policy initiatives. At your next dinner party just ask how many present would like to see society embark on a massive program of ecological tax reform geared to reducing consumption while maintaining quality of life. Even people who acknowledge the problem will become amazingly creative in generating reasons for defeating your proposal. For example, people who have never seemed particularly concerned about social inequity suddenly become worried about how the poor would fare faced with increasing prices, even though such a scheme could readily be designed (e.g., through income tax relief) to mitigate both existing inequity and future hardship.

In summary, it is our experience that systematically exploring the policy implications of Footprint analysis with focus groups or wider audiences can help raise to consciousness the contradiction between expressed levels of concern and the willingness to act while assisting researchers to identify the most important local barriers to sustainability. It also sheds light on how better to deliver difficult messages. This kind of guided social introspection is essential to assembling all the pieces in the sustainability puzzle.

16) Calculate your own Footprint

We would have liked to calculate our own Ecological Footprints but have been so busy writing this book that we have not been able to get down to it.... However, it is possible to estimate an individual Footprint and if you hurry, you can beat us to it! It's a bit tedious but not that difficult.

First you must keep a record of all your consumption-related expenditures and categorize them under headings such as housing, food, transportation,

Figure 3.18: By monitoring your personal or household consumption, you can calculate your own Footprint.

goods, and services.[40] It would be best to account for your consumption not only in dollars but also in litres, gallons, kilograms or other physical measures. To be thorough, you should also weigh and measure all the garbage (including recycling waste) that leaves the household. Your utility bills will tell you how much energy your house requires, and an accounting booklet in your car will help you keep track of gasoline consumed. Also, you should account for the services that you receive (such as health care or schooling) even if you do not pay its full costs.

Using the data and tables we have compiled for the average Canadian Footprint will help you to translate this consumption data into land areas. Additional documentation available from the UBC Task Force contains information about the embodied energy and resources of various consumption items. Unavoidably, some data will be missing, and you will have to rely on best guesses or on further research to fill these gaps. You will probably need a full year to account for all the variation in consumption related to holidays, seasonal heating and cooling requirements, and Christmas shopping. A reminder: be careful with the measurement units and time frames. For example, ecological productivity is usually recorded on an annual basis so for average monthly rates, you would divide by 12. Remember too that a Footprint of four hectares means that you consume the continuous production/assimilation by four hectares of ecologically productive land. In other words, your average monthly consumption corresponds to the average monthly output/input of your four hectare Footprint; your daily consumption averages your full four hectares' daily production.

Let us know how you do and what problems you encounter.

Figure 3.19: The eco-label of this newspaper could say: "Regular purchase and disposal of this product will claim just over 10% of your daily Earthshare or 2.5 hours of your daily fair share of global ecological output."

17) Eco-labelling: Is your product sustainable?

It might be surprising news, but there are no unsustainable products (excepting products that release non-degradable toxic substances such as radioactive materials). Sustainability is not about consumption *per se*, but about the *rates* of consumption. For example, it might be sustainable to operate a gas-guzzling Rolls Royce if it were shared among 20 friends, and maintained for a long time. On the other hand, it might be unsustainable for everybody to own an electric car. "Traditional" eco-labels tell us only whether a product is more ecologically benign than other similar products but nothing about the cumulative effects of mass consumption. Ecological Footprint analysis could improve product labelling by linking individual consumption to global ecological constraints.

Eco-labels could incorporate Ecological Footprinting by, for example, stat-

---

BOX 3.8: The Ecological Footprint of a Newspaper, Or, how much of your fair Earthshare (1.5 hectares) does your daily newspaper claim?

CALCULATIONS:
As a first approximation, the Ecological Footprint of a newspaper (assumed to weigh 0.3 kilograms) can be estimated by looking at two major resource inputs on which newsprint production depends: processing energy and fibres.

Energy requirements: It takes about 61 megajoules to produce one kilogram of paper. Therefore, the newspaper represents:

61 [MJ/kg] x 0.3 [kg] = 18.3 [MJ] embodied energy. With a yearly productivity of 100 GJ/ha/yr., or 150 GJ/yr. per Earthshare, 18.3 MJ corresponds to (8,760 [hours/yr.] x 18.3 [MJ] / 150,000 [MJ/yr.] =) *1.1 hours.*

Fibre requirements: Average wood fibre requirements for paper in Canada (in addition to the recycled fibres) amounts to 1.8 m³/t.

Fibre production in average forests is about 2.3 m³/ha/yr. The wood production of an Earthshare covered with forest would therefore be 1.5 x 2.3 = 3.5 [m³/yr.].
8,760 [hours/yr.] x (1.8 [m³/t] x 0.3 [kg newspaper] x 0.001 [kg/t]) / 3.5 [m³/yr.] = *1.4 hours.*

RESULT:
The consumption of a 300 gram newspaper occupies your fair Earthshare for (1.1 [hours] + 1.4 [hours] =) **2.5 hours.**

---

ing how much of a typical consumer's fair Earthshare would be occupied by average use of a certain good. Remember, a "fair Earthshare" refers to the area of ecologically productive land "available" *per capita* on Earth. In 1994, this was about 1.5 hectares. Living sustainably (and equitably) may be interpreted as using only those resource flows and waste assimilation capacities (natural income) that can be provided continuously by one's fair Earthshare. For example, the Rolls Royce will use about 5,000 gigajoules of energy over its life-cycle assuming it will run 500,000 kilometres. This corresponds to the production of about 50 hectares of ecologically productive land for an entire year. If you share the Rolls Royce with 19 other people and operate it for 20 years, it will claim about 8 percent of your 1.5 hectare Earthshare or almost two hours out of every 24 hours-worth of your personal ecological production.

This application departs a little from other Footprint applications: usually we estimate how much land a given consumption item appropriates. In this case, we start from a fixed land area (the fair Earthshare) and use EF analysis to determine how we must live to remain within personal — and ultimately global — carrying capacity. We cannot expand our fair Earthshare, we can only choose how much and how quickly we consume. Let's look at another example: your daily newspaper. The eco-label of the newspaper could say: "Regular purchase and disposal of this product will claim just over 10 percent of your Earthshare or 2.5 hours of your daily fair share of global ecological output" (see Box 3.8).

Similarly, research by the Wuppertal Institute in Germany into the energy and material costs of orange juice production shows that a daily litre of orange juice would occupy four to eight percent of your 1.5 hectare Earthshare (1–2 hours of your daily ecoproductivity) if produced in Brazil, or 26 to 30 percent of your Earthshare (7–8 hours of daily ecoproductivity) if produced on mechanized farms in Florida![41] At this latter rate, even a routine one glassful a day would take up 6–7 percent of your fair Earthshare.

So much for "fun with Footprints." We hope these 17 applications of Ecological Footprint analysis suffice to illustrate the scope and possibilities of this concept and to keep you interested in using it for your personal planning for a sustainable world.

Notes

1. Aubrey Diem, "Clearcutting British Columbia," *The Ecologist*, Vol.22 No.6, p.261-266, 1992. Mario Giampietro and David Pimentel, "Energy Analysis Models to Study the Biophysical Limits For Human Exploitation of Natural Processes." p.139-184, in C. Rossi and E. Tiezzi, (editors), *Ecological Physical Chemistry – Proceedings of an International Workshop*, Amsterdam: Elsevier, 1990.
2. Lester R. Brown, "Facing Food Insecurity" and Peter Weber, "Safeguarding Oceans" both in Worldwatch Institute, *State of the World*, NY: W.W.Norton, 1994. Carl Folke and Ann Marie Jansson, "The Emergence of an Ecological Economics Paradigm: Examples from Fisheries and Aquaculture," in U. Svendin and B. Aniansson, (editors), *Society and the Environment*, Dordrecht: Kluwer Academic Publisher, 1991. Michelle Hibler, *Our Common Bowl: Global Food Interdependencies*, Ottawa: International Development Research Centre (IDRC), 1992. (Yoshihiko Wada, PhD student in The University of British Columbia's School of Community and Regional Planning, is estimating Japan's terrestrial and marine Ecological Footprints including also products from the sea: "Assessing the Sustainability of Japan: The Ecological Footprint of an Average Japanese.")
3. The World Conservation Union, United Nations Environment Programme and the World Wide Fund for Nature, *Caring for the Earth: A Strategy for Living Sustainably*, Gland, Switzerland: IUCN, UNEP, and WWF, 1991.
4. William R. Catton Jr., *Overshoot: The Ecological Basis of Revolutionary Change*. Urbana: University of Illinois Press, 1980.
5. E. Mark Harmon, William K. Ferrell and Jerry F. Franklin, "Effects on the Carbon Storage of Conversion of Old Growth Forests to Young Forests," Science, Vol.247, p.699-702, 1990. Gregg Marland and Scott Marland, "Should We Store Carbon Trees?" *Water, Air and Soil Pollution*, Vol.64, p.181-195, 1992. Maria Wellisch, *MB*

Carbon Budget for the Alberni Region: Final Report, Vancouver: The Research and Development Department of MacMillan Bloedel Limited, 1992.

6. Mathis Wackernagel, "Ecological Footprint and Appropriated Carrying Capacity: A Tool for Planning Toward Sustainability," Unpublished PhD Thesis. Vancouver: University of British Columbia School of Community and Regional Planning (1994).

7. Yoshihiko Wada, "Biophysical Productivity Data for Ecological Footprint Analysis," Vancouver: Report to the UBC Task Force on Healthy and Sustainable Communities, 1994. New Zealand Forest Owner Association Inc., Forestry Facts and Figures 1994, Wellington, New Zealand: New Zealand Forest Owner Association in co-operation with the Ministry of Forestry, 1994.

8. Yoshihiko Wada, see above.

9. Salah El Serafy, "The Proper Calculation of Income from Depletable Natural Resources," in Ernst Lutz and Salah Serafy, *Environmental Resource Accounting and Their Relevance to the Management of Sustainable Income*, Washington DC.: The World Bank, 1988.

10. Vaclav Smil, *General Energetics: Energy in the Biosphere and Civilization*, NY: John Wiley, 1991. David Pimentel, "Achieving a Secure Energy Future: Environmental and Economic Consequences," *Ecological Economics*, Vol.9 No.3, p.201-219, 1994. Michael Narodoslawsky and Christian Krotscheck, "The Sustainable Process Index Case Study: The Synthesis of Ethanol from Sugar Beet," Technische Universität Graz, Austria: Institut für Verfahrenstechnik, 1993.

11. If not otherwise indicated, all the data in this chapter stems from the World Resources Institute's bi-annual *World Resources Report* (NY: Oxford University Press).

12. 1,111,000,000 [Gigajoules per year] / 1,000 [Gigajoules per hectare per year] / 27,000,000 [Canadians] = 0.04 hectares per Canadian.

13. Calculations by Mathis Wackernagel and Yoshihiko Wada (building on Carl-Jochen Winter and Joachim Nitsch's *Hydrogen as an Energy Carrier: Technologies, Systems, Economy* [Berlin: Springer Verlag, 1988]) suggest a photovoltaic productivity of 100 to 500 Gj/ha/yr. Michael Narodoslawsky, Christian Krotscheck and Jan Sage ("The Sustainable Process Index (SPI): A Measure for Process Industries," Technische Universität Graz, Austria: Institut für Verfahrenstechnik, 1993) arrive at a productivity of 430 Gj/ha/yr. David Pimentel et al. (see above) list 1,200 Gj/ha/yr for photovoltaics. All the following renewable energy estimates build stem from these sources as well as from Vaclav Smil (see above).

14. In effect, we are already doing this in the case of intensive high-input agriculture. Output is more dependent on the fossil energy and material "subsidies" than it is on the remaining natural productivity the land.

15. Barney Foran,CSIRO, Australia, Division of Wildlife & Ecology, personal communication, November 1994.

16. World Resources Institute, *World Resources*, NY: Oxford University Press, 1992. Sandra Postel and John Ryan, "Reforming Forestry," in Worldwatch Institute, *State of the World*, NY: W.W. Norton, 1991.

17. Food and Agriculture Organization of the United Nations (FAO), *FAO Yearbook: Production*, Vol. 43, Rome: FAO, 1990. World Resources Institute, *World Resources: Data Base Diskette*, Washington, DC: World Resources Institute, 1992. Statistics Canada data.

18. Rather than the 250 Gigajoules per capita per year in Canada, Statistics Canada claims only 234. To keep the calculations conservative, this latter figure used for estimating the Canadian Footprint.

19. See Robert Smith, "Canadian Greenhouse Gas Emissions: An Imput-Output Study," in *Environmental Perspectives 1993: Studies and Statistics*. Ottawa: Statistics Canada (Catalogue 11-528E Occasional).

20. For Canada, the average mature forest contains 163 m³/ha of useful timber. Assuming a harvest rotation period of 70 years for temperate forests, this would result in a productivity of about 2.3 m³/ha/yr – similar to typical figures for the "Annual Allowable Cut" in public forests. Data compiled by Gregg and Scott Marland ("Should We Store Carbon in Trees?" Water, Air and Soil Pollution, Vol.64, p.181-195, 1992) suggest that the world timber productivity would average 4.1 m3/ha/yr. This is calculated from boreal productivities of 2.3 m³/ha/yr (corresponding to 33 percent of the global forest area), 3.3 m³/ha/yr for temperate forests (25 percent of the area) and 6 m³/ha/yr for tropical forests (42 percent of the area). However, the yield for tropical forests is speculative so the reliability of the global estimate is questionable. Another way to calculate average timber productivities is through carbon accumulation data. Yoshihiko Wada's survey of the literature suggests a carbon absorption rate of 1.8 t/ha/yr ("Biophysical Productivity Data for Ecological Footprint Analysis," Vancouver: Report to the UBC Task Force on Healthy and Sustainable Communities, 1994). This corresponds to about 4 t/ha/yr dry biomass of which a maximum of 25 percent might be merchantable timber. With an average density of approximately 0.5 t/m³, this would result in about (4 [t/ha/yr] x .25/ 0.5 [t/m³] =) 2 m³/ha/yr.

21. Maria Buitenkamp, Henk Venner and Theo Wams, (editors), *Action Plan Sustainable Netherlands*, Amsterdam: Dutch Friends of the Earth, 1993. In this report, they propose the Environmental Space concept which is complementary to the Ecological Footprint (see Box 2.5).

22. Again we recognize the relentlessly anthropocentric tone of this discussion. This is not because we discount the intrinsic rights and values of other species. It simply acknowledges the ecological reality that humankind is already the dominant species in all the world's ecosystems and recognizes that, at present, economic narcissism dominates human attitudes.

23. On grounds that some of the functions of the suggested wilderness reserve could be assumed by biodiverse carbon sink forests.

24. International Institute for Environment & Development, *Citizen Action to Lighten Britain's Ecological Footprints*, London: International Institute for Environment & Development, 1995.

25. National Institute for Public Health and Environmental Protection (RIVM), *National Environmental Outlook 2*, 1990-2010, Bilthoven, Netherlands: RIVM, 1992.

26. Figures based on World Resources Institute data and information from the above-mentioned Action Plan Sustainable Netherlands. Please note that, in this example, the ecologically productive land areas (Column a) are not adjusted for productivity. World average productivity is assumed. This does not weaken the argument: even if local productivity was double the world average (which is unlikely), the deficits would still be significantly larger than the available land areas — still with the

exception of only Australia and Canada.

27. Rod Simpson, Katherine Gasche and Shannon Rutherford, *Estimating the Ecological Footprint of the South-East Queensland Region of Australia (draft report)*. Brisbane: Faculty of Environmental Studies, Griffith University, 1995.

28. Footprint sizes are inferred from studies by Ingo Neumann from Trier University, Germany; Dieter Zürcher from Infras Consulting, Switzerland; Rod Simpson, Katherine Gasche and Shannon Rutherford of Griffith University, Australia; and our own analysis using World Resources Institute (1992) data.

29. Data from the World Resources Institute.

30. If ecological productivity is greater than assumed, then this deficit would be somewhat smaller.

31. William M. Alexander, "Humans Sharing the Bounty of the Earth: Hopeful Lessons from Kerala," paper prepared for the International Congress on Kerala Studies in Thiruvanathapurum, Kerala, 1994.

32. Thomas Homer-Dixon, Jeffrey H. Boutwell and George W. Rathjens, "Environmental Change and Violent Conflicts," *Scientific American*, p.38-45, February 1993. Clive Ponting, *A Green History of the World: The Environment and the Collapse of Great Civilizations*, NY: St. Martin's Press, 1992.

31. Nick Robins at the International Institute for Environment and Development in the UK is developing a similar study to analyze the impact of international trade and its implications for national policy.

33. Cited in Robert Goodland and Herman E. Daly, "Why Northern Income Growth is not the Solution to Southern Poverty," *Ecological Economics* Vol. 8, No. 2, p.85-101, 1993.

34. World Council of Churches, "Accelerated Climate Change: Signs of Peril, Test of Faith," Geneva, 1992

35. In Mark Roseland, *Toward Sustainable Communities*, Ottawa: National Round Table on the Environment and the Economy, 1992. Free copies can be obtained from the Round Table at tel. (613) 992-7189.

36. Yoshihiko Wada, *The Appropriated Carrying Capacity of Tomato Production: The Ecological Footprint of Hydroponic Greenhouse versus Mechanized Open Field Operations*. Vancouver: M.A. Thesis at the UBC School of Community and Regional Planning, 1993.

37. For example: Julian Griggs, Tim Turner, and Mathis Wackernagel, *Connections: Towards a Sustainable Future (A Four-Day Program on Sustainability)*, Draft Program Guide, Gibsons, B.C.: Sea to Sky Outdoor School for Environmental Education, 1993. ESSA Technologies Ltd., *Teacher's Guide to the State of the Environment Report for British Columbia*, Victoria, B.C.: Ministry of Environment, Lands and Parks, 1994. Jim Wiese, *Energy Education – Module 4: Conservation Potential*, (section on Ecological and Energy Footprints), Vancouver: BC Hydro, 1995.

38. Colin Duffield, *Putting the Ecological Footprint in Print: Applications of the Appropriated Carrying Capacity Concept to SOE Reporting*, Ottawa: Report to Strategic Planning and Analysis of SOE Reporting, 1993.

39. Peat Marwick Stevenson & Kellogg, *Sustainability Indicators Methodology, Report prepared for the Fraser Basin Management Board in Vancouver*, 1993, and; *1993 Environmental SCAN: Evaluating Our Progress toward Sustainability*, Ottawa: Canadian Council of Ministers of the Environment, 1993.

40. A useful accounting framework for personal consumption is explained in Joe Dominguez and Vicki Robin's book, *Your Money or Your Life* (NY: Viking Penguin, 1992).
41. Friedrich Schmidt-Bleek, *Wieviel Umwelt braucht der Mensch: MIPS - das Mass für ökologisches Wirtschaften, (How Much Environment Do People Need? MIPS: The Measure for Managing Ecological Economies)*, Basel and Boston: Birkhäuser, 1993.

4

THE SEARCH FOR
SUSTAINABILITY
STRATEGIES

The global Ecological Footprint demonstrates that we no longer live in a world with abundant, unused ecological capacity. Few ecosystems remain whose production has not yet been modified to serve human demands and all are being used in some way that sustains human activities. In fact, evidence indicates that the human load has substantially exceeded long-term global carrying capacity. This overshoot turns humankind into a "catch-22" victim of its own making. More material growth, at least in the poor countries, seems essential for socioeconomic sustainability, yet any global increase in material throughput is ecologically unsustainable.

Acknowledging this sustainability challenge is psychologically disturbing. It implies that the human race cannot safely continue on its current path, a path which by many modern values has been stunningly successful in improving human welfare. Profound changes will be required. In particular, we wealthy members of the human family — the *average* residents of the industrial world — face a discomforting moral dilemma: while we consume on average three times our fair share of sustainable global output, the basic needs of the world's billion plus chronically poor are not being met even today. Meanwhile, just satisfying current aggregate demand is undermining nature's capacity to meet the needs of future generations. If we rely on conventional economic strategies and technologies to fix these problems, the additional material growth, particularly in the high-income countries, would appropriate even more carrying capacity for the rich, thus reducing the ecological space available to the poor. It seems that conventional strategies are both ecologically dangerous and morally questionable. To the extent that we can create room for growth, it should be allocated to the Third World.

The challenges before us are unprecedented: How can we decrease humanity's total ecological impact while providing adequately for everyone's needs? Who should be asked to reduce their Ecological Footprints, and who may increase theirs to meet basic needs? What would convince anyone of the need

Figure 4.1: Current strategies detract from sustainability, undermining both ecological and moral integrity. Any available space for material growth should be allocated to those whose basic needs are not being met.

to reduce his/her Ecological Footprint? What social, institutional and technological mechanisms are available to help us do so? In short, how can we devise
a social contract to build a sustainable society, one that relieves the weakest
from bearing the greatest burden of the sustainability crisis, and ensures a
satisfying existence for all?

Questioning Conventional Strategies

Most of us are sold on conventional strategies and the promise that they are
compatible with sustainability. The rhetoric is familiar: some people advocate
"sustainable growth" and promote freer trade as the way to achieve it. The
debates over the North American Free Trade Agreement (NAFTA) and the even
more encompassing General Agreement on Tariffs and Trade (the GATT,
recently reborn as the World Trade Organization) are familiar to most of us.
Others emphasize technological solutions such as "zero-emission" cars and the
potential of a new efficiency revolution to create the ecological space for
continued GDP growth.

In fact, conventional analysts, including many contributors to *Our Common
Future*, argue that trade and technology actually extend any ecological limits.
This is a partial misconception. Even in the best of circumstances, technological
innovation does not increase carrying capacity *per se* but only the efficiency of
resource use. In theory, shifting to more energy- and material-efficient technologies should enable a defined environment to support a given population
at a higher material standard, or a higher population at the same material
standard. However, while this seems to increase carrying capacity it actually
only holds total human load constant in the vicinity of carrying capacity. The
latter is unchanged and ultimately still limiting.

Moreover, in practice, efficiency gains and current incentives often work
directly and indirectly *against* resource conservation. Many factors contribute
to this counter-intuitive result, including the price and income effects of
technological savings. Improved energy or material efficiency may enable
firms to raise wages, increase dividends or lower prices, which leads to
increased net consumption by workers, shareholders or consumers respec-
tively. Similarly, technology-induced savings by individuals are usually
redirected to alternative forms of consumption, cancelling some or all of the
initial benefit to the environment. To the extent that such mechanisms contribute to increased consumption and accelerated resource depletion, efficiency
gains indirectly increase the affected population's Ecological Footprint, adding
to the burden on the planet's limited carrying capacity (see Box 4.1).

Worse, many industrial technologies contribute directly to larger Ecological
Footprints, while creating the illusion of shrinking them. This is particularly
evident in agriculture, forestry and mining where technology boosts the short-

128 OUR ECOLOGICAL FOOTPRINT

BOX 4.1: Will Efficiency Gains Save Resources?[1]

Many economists and environmentalists believe that advances in technological efficiency are a potential panacea for the sustainability crisis. This follows from Buckminster Fuller's reasoning of "doing more with less" and contains the hidden assumption that efficiency gains automatically lead to resource savings and reduced consumption. For example, industrialist Stefan Schmidheiny lauds the 50 percent energy efficiency gains by the chemical industry in recent decades, forgetting that chemical production has doubled in the same period. Even *Our Common Future* was devoted to what Wolfgang Sachs calls "the gospel of global efficiency." However, as effective as these efficiency strategies might seem on the micro-scale, decreasing the ratio between input and output does not necessarily lead to lower resource use. On the contrary, technological efficiency may actually lead to increased net consumption of resources.

Various authors have recognized the resultant dilemma. *Limits to Growth* pointed out in 1972 that a doubling of agricultural productivity accompanied by continued economic expansion would extend food limits by only 20 years and leave us with a more intractable problem. Lester Brown from the Worldwatch Institute observes that "...continuing growth in material consumption — the number of cars and air conditioners, the amount of paper used, and the like — will eventually overwhelm gains from efficiency, causing total resource use (and all the corresponding environmental damage) to rise...." Real data confirm this speculation: in the U.S., despite the increasing fuel efficiency of cars, aggregate fuel consumption is on the rise. Similarly, as *The Ecologist* notes, while energy use per dollar Gross National Product (GNP) decreased by 23 percent in the Western industrialized countries between 1973 and 1987, total annual energy consumption actually increased by 15 percent over the same time span.

On the micro-level: Improved energy or material efficiency may enable firms to raise wages, increase dividends or lower prices, all of which lead to increased net consumption. Economists call these wage and price effects the "rebound effect." Similarly, technology-induced savings by individuals are usually redirected to other forms of consumption, cancelling some of the initial gain. As ecologist and energy analyst Bruce Hannon explains, "...the [environmentally conscious] traveller who [switches] from urban bus to bicycle would save energy (and dollars) at the rate of 51,000 BTU per dollar. If he were not careful to spend his dollar savings on an item of personal consumption which had an energy intensity greater than 51,000 BTU per dollar then his shift to bicycle would have been in vain...."

On the meso (or regional) level: Typically, industrialized countries import much of their energy, mostly in the form of fossil energy. Such imports weaken local economies through the loss of direct spending (leakage) and the loss of that spending's re-spending (the multiplier effect). By contrast, the money for both the energy-saving equipment as well as the moneys saved through improved energy efficiency will most likely be spent locally, thus stimulating the local economy.

On the macro-level: According to economist Paul Samuelson, technical innovations or efficiency gains account for 75 percent of GNP growth, thereby contributing to increased aggregate resource throughput. Analyzing the effects of efficiency gains, economist Harry Sanders concludes that "...energy efficiency gains can increase energy consumption by two means: by making energy appear effectively cheaper than other inputs; and by increasing economic growth, which pulls up energy use...." Other studies reject the claim that GNP and energy consumption have ever been decoupled in industrialized countries. Energy analyst Robert Kaufmann concludes that substitution and technical change have had relatively little effect on the amount of energy used to produce one dollar of inflation-adjusted GNP in France, Germany, Japan and the U.K. since the Second World War. This implies that the link between economic activity (measured in GNP) and energy use is stronger than believed by most neoclassical economists.

In general, it seems that technical efficiency gains that produce increased returns to capital will attract investment and ripple through the economy. As economist Stanley Jevons observed in 1865 in *The Coal Question*: "...the progress of any branch of manufacture excites a new activity in most other branches and leads indirectly, if not directly, to increased inroads upon our seams of coal...." In short, profitable efficiency gains — and these are the ones that get implemented — contribute to upward-trending expectations of returns to capital and higher investment in efficient firms. This induces the competitive spread of the efficient technologies to other firms and sectors, which may well increase total demand for resources.

Ironically then, it is precisely the economic gains from improved technical efficiency that increase the rate of resource throughput. Micro-economic reality demands that these efficiency gains be used to short-term economic advantage. Far from conserving natural capital or decreasing Ecological Footprints, this leads to competitively accelerated increases in consumption. In a globally interlinked economy, the question then becomes: Can we afford cost-saving energy efficiency? The answer is "yes" only if efficiency gains are removed at source from further economic circulation.

Ideally, efficiency savings should be captured for investment in natural capital rehabilitation. This can only be achieved in the relatively short term through the institution of resource depletion taxes, marketable resource quotas, and other elements of ecological tax reform (including reductions in income taxes and other penalties on labor). If we don't implement policies that will force us to do more with less now, we may be forced by nature later to do the same (or even less) with less later!

term "harvest" (as demonstrated by our comparative assessment of alternative methods of producing tomatoes). We are often led to believe that these elevated yields imply "improved" productivity of natural capital when, in fact, they stem from the liquidation of capital stocks. Thus, intensive agriculture may be more "productive" than low-input farming in the short term, but depends on external energy subsidies and increased rates of soil and water depletion. The net effect is that society becomes increasingly dependent on enhanced flows of non-renewable resources while actually reducing the long-term (and renewable) carrying capacity of its environment. The potential consequences of this reality in times of stress were experienced cruelly in Cuba, where agricultural production collapsed with the shortage of fossil fuel in the early 1990s.

Of course, properly employed, technology can play a major contribution to a more sustainable society. As shown in Chapter 3, solar technology can massively reduce our energy Footprint. Energy saving devices should therefore be welcomed as long as these savings are not simply diverted to other forms of consumption. In contrast, proposals such as that of switching to so-called "zero-emission" cars seem more problematic. Attempts to reduce air pollution in urban areas are commendable, but eliminating such urban pollutants as nitrogen oxides and unburned hydrocarbons from car exhausts ignores CO_2 emissions (a major contributor to potential climate change) and the fact that end-of-the-pipe emissions from cars are only a fraction of their total ecological impacts. Even zero-emission cars pollute and use resources when being built, operated and junked. And, if zero-emission cars continue to encourage more autos and auto use — the engines of urban sprawl — they may help in other ways to expand our Ecological Footprints.

The apparent gains in carrying capacity that come from trade are also illusory. While commodity trade may release a local population from resource constraints imposed by its own home territory, this merely displaces parts of that population's total environmental load to distant export regions. Whenever a local population is able to import carrying capacity, that population or its economic activities invariably expand. However, this does not represent a net gain in carrying capacity; the expansion of import regions is accompanied by reduced load-bearing capacity in export regions. Meanwhile, the trade-induced increase in human numbers and consumption increases the aggregate load of people on the ecosphere.

Even so, it is possible to conceive of ecologically sound and balanced trade. If each nation were to export only true surpluses — output in excess of local consumption whose export would not deplete self-producing natural capital stocks — then the net effect would be an ecological steady-state and global stability. However, like technology, unregulated trade may well contribute to *reducing* the long-term carrying capacity of both partners in an exchange. Access to cheap imports (e.g., food or timber) both lowers the incentive for

Figure 4.2: Have Efficiency Gains Reduced Our Footprints? *"Increase efficiency!"* is a popular rallying cry in efforts to reduce our impact on the planet. Does it really work? Not always. Economist Stanley Jevons warned us a century ago that it is a confusion of ideas to suppose that the economical use of resources is equivalent to diminished total consumption.

Figure 4.3: Ecological Trade Imbalance. Monetary analyses reveal accumulations of money wealth but tell us little about the material flows that make them possible. On the other hand, physical models such as Ecological Footprint analysis would show the increasing dependence of industrial nations on imported flows as well as the extent to which resource-based economies trade off low-priced natural capital (i.e., carrying capacity) to buy expensive manufactured goods and services from value-adding economies. Can we continue to ignore the long-term consequences of such ecologically imbalanced and globally damaging trade?

importers to conserve their own local natural capital stocks (e.g., agricultural land or forests) and may result in the competitive depletion of the exporters' assets as well (competition for export markets lowers prices, reducing any surplus available for natural capital maintenance).

Let's pursue this last point for a moment. It seems that, in today's world, urbanization, globalization and trade combine to reduce corrective feedback on local populations. With access to global resources, urban populations everywhere are seemingly immune to the consequences of locally unsustainable land and resource management practices — at least for a few decades. In effect, modernization alienates us spatially and psychologically from the land. The citizens of the industrial world suffer from a collective ecological blindness that reduces their collective sense of "connectedness" to the ecosystems that sustain them.

Meanwhile, trade has obviously been a major factor in raising Gross World Product in the post-war period. Less obviously, it has also become a major mechanism by which the rich (inadvertently?) appropriate much of the world's carrying capacity and extend their own Ecological Footprints. However, as the total human load on the ecosphere increases and the depletion of natural capital accelerates, trade reduces the ecological safety net for all.

In this light, far from overcoming the ecological limits to material growth, expanding trade assists humanity to overshoot dangerously the long-term carrying capacity of the planet. International terms of trade should therefore be re-examined to ensure that they are equitable, socially constructive and confined to true ecological surpluses. In general, trade should be managed to prevent further depletion of renewable natural capital and so that the benefits of ecologically safe growth flow to those who need them most. To begin, export taxes or import tariffs should be permitted to ensure that prices reflect at least known ecological externalities.

Ecological Footprint analysis could be used to facilitate implementation of such a trade regime. The establishment of a set of regional ecological (physical) accounts could assist countries or bioregions to compute their true ecological loads on the ecosphere and to monitor their ecological trade balances. Such accounts would enable the world community to ensure that global flows do not exceed sustainable natural income and that humanity lives within global carrying capacity.

Finally, we wish to emphasize that none of these comments are to be taken as arguments against technology or trade *per se* — we acknowledge and welcome the benefits that both have brought to our lives and each has an important role in any sustainable future. Rather, our point here is to emphasize that conventional assumptions about trade and technology should be carefully re-examined in light of carrying capacity considerations and that certain conditions must be satisfied before either can contribute to ecological sustainability.

The Process of Developing Sustainability

Hundreds of books, government documents and NGO brochures propose myriad imaginative strategies for dealing with the sustainability problem. The development of sustainability hinges not only on what is done, but on how we go about doing it. Even the best ideas will not bear fruit if they do not fit their context or if they lack the support of the affected people. Before discussing concrete ideas and visions about sustainability, we need to think first about processes for effective action. The following concepts have guided our community work with the UBC Task Force on Healthy and Sustainable Communities.

The two sustainability poles: ecological stability and human quality of life

The necessary conditions for developing sustainability are straightforward, at least on the surface. In the industrialized countries our strategy must simultaneously reduce our Ecological Footprints (the ecological imperative) while securing a satisfactory quality of life for all (the socioeconomic imperative). These are the main poles of tension between which planning for sustainability operates.

Ecological sustainability is the conceptually simple part of the sustainability concept: while there is considerable debate over where exactly the limits are, there is general consensus that we must learn to live together within the means of nature. Failing to live on sustainable natural income will put human(e) survival at risk. Ecological Footprint analysis can be used to measure progress toward achieving this requirement. Ecological Footprint applications and similar analyses not only demonstrate ecological limits, but also suggest ways to translate global constraints into specific action at smaller scales such as the regional, municipal or individual levels. EF analysis estimates how much of nature's productivity is required to maintain a given lifestyle and helps determine whether nature can actually provide these flows over the long run. It also provides a test to determine whether given policy choices will increase or decrease our ecological demands. In other words, it is a tool that can help assess the long-term prospects for humane life on this planet.

Socioeconomic sustainability is a more difficult and potentially contentious concept. In economic terms, the minimal goal would be for everyone to be able to attain a material standard sufficient for them to enjoy an emotionally and spiritually satisfying life. At present, of course, many people in the industrial world have long surpassed any such "sufficiency" standard. Indeed, these people enjoy extremely high material standards and, at least initially, would want to maintain their consumer lifestyles. Meanwhile, over a billion people on Earth are unable to satisfy even their basic material needs. The problem is, how can we reconcile the economic disparity between the rich and the poor at the limits of ecological stability in a socially just and politically acceptable manner? Social inequity and the material disparity that accompanies it are therefore at the center of the sustainability debate. The question of "who gets what (and how)" raises the spectre of potential conflict both within and between nations. The need for distributive justice and the associated latent conflict is the most scary and politically taxing part of the sustainability equation.

The fact is that civil strife and international conflicts can be induced or worsened by ecological overshoot if there is already marked socioeconomic inequality. Many scholars warn that excessive demands on nature that produce resource shortages and ecosystem collapse will not only fuel local strife, but also threaten political instability at the global level. Most immediately, many people see their interests and aspirations undermined by other people's claims and seem instinctively to dread the changes this implies. (Psychological studies show that possible gains have to outweigh threatened losses threefold before people are willing to accept a change in their lives. For example, think how much better public transportation would have to be before North Americans would give up their cars in the city, even though the latter often eat a quarter of their budgets.)

Naturally, everybody wants a secure and fulfilling life. Achieving this hinges on both material and social prerequisites such as adequate food, shelter, clothing, health care, education, and being an accepted part of a supportive and friendly community. To address social conflict and to make progress toward socioeconomic sustainability, we must better understand what "improving quality of life" means. First, we need to recognize that changes in, or comparisons between, the quality of life of communities can be only partially observed from the outside — for the complete picture, communities have to evaluate their situation themselves. Special participatory planning processes can assist people to reflect more systematically upon their community's quality of life and to analyze the potential impacts of alternative policy choices. A classification of needs that might be useful in these circumstances has been suggested by economist Manfred Max-Neef who observed similar needs in all cultures and all historical circumstances. He identifies these needs as: permanence (or continuity); protection; affection; understanding; participation; leisure; creation; identity (or finding meaning); and freedom.[2] Using such tools can help affected communities work with their planners to determine whether socioeconomic requirements for sustainability are being met and how gaps should be addressed.

Will it be possible simultaneously to achieve both the ecological and socioeconomic conditions for sustainability? Satisfying ever more people's desire for a good life in a world of increasing material inequity and declining resources is a demanding challenge. It is the more complicated because ecological sustainability is perceived as a long-term global concern while prevailing socioeconomic conditions affect everyone now in their own communities. Given the human tendency to discount the future, it is no surprise that the latter concern often overshadows the former. Acknowledging the frequent neglect of long-term ecological security in favor of immediate quality of life issues therefore becomes a starting point for developing sustainability. How can we reconcile these mutually dependent conditions? Can everyone enjoy the "good life" without jeopardizing "the future for our children"? Clearly, society must attempt to understand how quality of life decisions today affect the ecological security of future generations.

Win-win solutions

In a culture in which much is good, more is better, and too much may not be enough it may not seem possible to improve the quality of life while reducing our Ecological Footprints. However, Vicki Robin of the New Road Map Foundation argues that these trends are actually complementary and there is plenty of academic research to support her claim.[3] Once material sufficiency is secured, people's happiness is no longer correlated with national

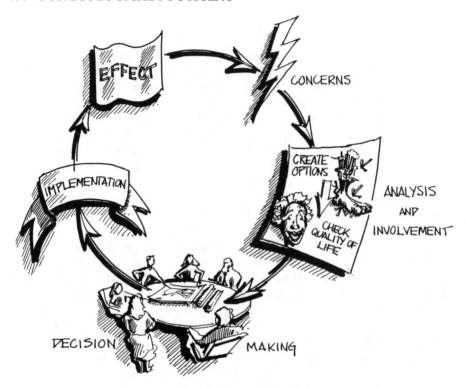

Figure 4.4: We often need several circuits 'round the cycle of change to translate community concerns into effective action.

or personal income. It seems that "more is better" is an inherently frustrating game promoted by the fallacy of confusing quantity of things with quality of life. It turns out that the best things in life are not "things." In fact, having fewer possessions does not need to deprive us, but can be liberating. (Of course, we already know that, but do we act as if we did?) True fulfilment comes from being with others and contributing to their lives, rather than from taking and withdrawing. Thousands of people have discovered that there may be personal benefits from down-scaling, such as living debt-free, having more time to live, and enjoying greater security. In effect, one can become more independent of "the system" — reducing our material wants can help us get out of the rat-race and off the "dreadmill." The trick is to focus our lives on maximizing fulfilment rather than income.[3]

Liberation comes from realizing that finding fulfilment comes not from having more but from needing less. This frees our time and our life energy — most precious resources — for doing things that matter to us. In industrialized countries we particularly need the time to strengthen and revitalize our communities. Sociologist Amitai Etzioni calls for a new Spirit of Community based on the shared values and mutual understanding required to restore civil society. His communitarian program starts from the observation that the proliferation of individual rights threatens community needs, and that a new balance has to be found to create a more co-operative and responsible social contract in which rights and responsibilities go together. Indeed, perhaps the most critical social condition for sustainability is a shared commitment to community cohesion (both global and local) and a sense of collective responsibility for the future.

The cycle of change in decision-making

Pushing the supposed "moral superiority" of sustainability will not make it happen. In today's fragmented and competitive world, playing on people's moral duty and feelings of guilt produces only resentment, not long-lasting transformation. Sustainability will remain a hard sell until we can show that people have more to gain than to lose by changing their ways. Change flows from necessity, hope, realizable aspirations and joy, not shame and blame.

Framing the potential conflicts that exist between maintaining quality of life today and ensuring ecological stability in the future is an essential element in the decision-making process for sustainability. This is an iterative rather than linear planning process. That is, it involves repeated cycles of learning (often by trial and error) and reconsideration, which gradually move people's beliefs and transform community concerns into action (see Figure 4.4).

The process may be initiated when a particular interest group or a municipality decides to achieve a particular purpose or react to a common concern. The initial phase includes acknowledging the need for a collective response, identifying potential conflicts and trade-offs, and determining ways to involve the public. To begin serious planning, the community must clarify its objectives, set priorities, and identify policy options for achieving them. Once the community has decided on a course of action and has begun to implement the agreed policies (i.e., "the plan"), it is important to monitor progress relative to the initial objectives. A generally perceived need to revise the plan or for further change initiates the cycle again.

The key to success in iterative planning is to develop clear policy options and a community-wide understanding of the decision-making process and criteria to be used in selecting among these options. Transparent processes enable healthy debate and constructive dialogue. Without feedback between decision-makers and affected interests, and without a broad understanding of the constraints and options, the best strategies are doomed to fail.

Ecological Footprint analysis can assist the development of policy options for sustainability. For example, once a community accepts the ecological imperative for change, it may wish to generate options to reduce its collective consumption. EF analysis can then be used to compare specific policy options

Figure 4.5: The Boiled Frog Syndrome. A frog placed in slowly heating water will not notice the gradual but eventually lethal trend.

and plans to determine which will move us most quickly toward sustainability. The challenge is to select those options that reduce the community Footprint while securing peoples' quality of life. Keep in mind: the community will accept only those options that are likely to produce greater personal security and community livability than would result if the options were not adopted.

Three uphill battles to achieve sustainability

We need to recognize that achieving sustainability will require fighting three uphill battles. We can characterize these as: the boiled frog syndrome, mental apartheid, and the tragedy of the commons.[4] First, our reductionist propensity to focus on mere symptoms of problems or on individual events detracts from our seeing the whole. We end up ignoring — or at least failing to anticipate — the cumulative effects of individual events. Neurologist Robert Ornstein and biologist Paul Ehrlich believe that our focus on isolated and immediate incidents is linked to the way the human brain functions: slow changes, long-term implications, and multiple connections cannot easily be perceived. This can be likened to the "boiled frog syndrome." Ornstein and Ehrlich explain that "...frogs placed in a pan of water that is slowly heated will be unable to detect the gradual but deadly trend.... Like the frogs, many people seem unable to detect the gradual but lethal trend in which population and economic growth threaten to boil civilization...."

If we do not wake up to the slow but steady deterioration of the planet, we will ultimately become victims of the "tyranny of now." Society's penchant for trading off the ecosphere in tiny bits to satisfy immediate wants is the ecological equivalent of the fire under the frog's pan. Worse, the incremental expansion of the human enterprise, of manufactured capital infrastructure catering mainly to today's already affluent, overwhelms others' efforts to live within the means of nature.[5] We hope that the graphic clarity of the Ecological Footprint, by showing all at once how much ecosphere we have *already* traded off, may wake the majority from the consumption-induced lethargy of our material age.

We also seem plagued by a form of mental apartheid that has erected an imposing psychological barrier between modern humans and the rest of reality. This perceptual dualism is clearly embedded in our language (which is itself a map of how we see the world). For example, the very term "environment" separates the really important stuff "in here" from everything else "out there." Our exemptionist attitude is also evident in the way we resist the notion that humankind is an integral part of nature, that we are just one of many millions of species occupying this planet.

Although it also has much earlier roots, the characteristic dualism of our techno-scientific culture is commonly associated with enlightenment philosopher René Descartes, who divided all of reality into separate realms of mind and matter. The former has become associated with humans and their works and the latter with everything else. Descartes supposed the whole of the external material world existed to be known, manipulated and exploited by people. This artificial mind-body split is clearly dysfunctional at the ecological limits of a finite world; our mental apartheid must be broken down. Sustainability requires a profound sense that the fate of the ecosphere is the fate of humankind — we do not *have* a body, we *are* a body; we are not surrounded by an "environment," we are an intimate part of the ecosphere. Again, the Ecological Footprint approach can help to restore our understanding that we are embedded in nature — unlike most mainstream "environmental" analyses, it does not show the impact of people *on* nature, but rather the dominant role of humans *in* nature.

The third behavioral syndrome that inhibits us from acting sustainably is the so-called "tragedy of the commons" (more accurately, "the tragedy of open access"). Ecologist Garrett Hardin reiterated Aristotle's wisdom that "...what is common to the greatest number gets the least amount of care..." and emphasized its tragic social implications. In general, this problem emerges whenever the benefits to an individual of (over-)exploiting an open-access resource exceed that individual's share of the resulting damage costs. Hardin compared an individual shepherd's gain from increasing his/her herd size on an open pasture to the same shepherd's share of the costs of doing so. Since the net benefits will always seem greater to the individual shepherd, each has a continuing incentive to add more animals to the pasture, eventually destroying it for all. Even if one good shepherd recognizes the imminent tragedy, there is no incentive for him/her to exercise personal restraint — someone else will simply fill the void.

This tragic mechanism is a major driver of the downward global ecological spiral (the 1995 "groundfish war" between Canada and the European Community in the North Atlantic is a recent visible example). Hardin advocated resolving the tragedy through social contracts to govern common-pool resources. What we need, in his words, is "...mutual coercion, mutually agreed upon...." This approach (which, in fact, describes ancient commons regimes) is consistent with the communitarian approach advocated by Etzioni. Ecological Footprint analysis can also contribute to our understanding of the open access problem. It shows the extent to which consumption today has already appropriated the output of the so-called "global commons."

There are, of course, many other psychological quirks — denial, dogmas and taboos — that cause us to maintain contradictory aspirations and inspire us to urge other people to adopt behavior that we are not willing to adopt ourselves. Clearly, we need better tools of all kinds to help us understand our role in the ecosphere, to make our decision processes more visible, and to clarify the trade-offs among the options before us.

Sketching a Vision for a Sustainable Society

Municipalities today are under pressure to deliver more services with fewer resources. Similarly, national governments are being weakened by soaring debt payments. In these circumstances, conventional economic development initiatives look the more attractive to all levels of government. At the limit to carrying capacity, however, growth for the sake of additional revenues is both an economic and an ecological negative-sum game. How can we move from today's unsustainable lifestyles to a more harmonious relationship with nature in our communities? As noted, local planning offers many possibilities — changes in transportation and land-use patterns alone at the municipal level can significantly reduce resource consumption and, at the same time, improve local quality of life. Moreover, since these new policies influence housing and commuting patterns, and not the economy directly, they are unlikely to endanger local competitiveness. If anything, they should reduce local land and transportation costs, increasing the comparative advantage of the relevant municipalities.

We have shown that four to five hectares of ecologically productive land are needed to support the average North American. This is orders of magnitude more land than is available per capita within typical urban municipalities, or even within their watersheds, and provides a measure of the local ecological deficit. It also demonstrates the great leverage inherent in firm action at the municipal level: a mere five percent cutback in resource input and waste generation in a region where the Ecological Footprint exceeds carrying capacity by a factor of 20 (as in the case of the Lower Fraser Basin in B.C.) will reduce the Ecological Footprint by an area the size of the entire region!

Of course, the positive effects of more sustainable urban design would be greatly enhanced if people also changed their behavior and lifestyles. For example, EF analysis suggests that we should focus more on living locally than on consuming globally. In many places we could still dwell comfortably on the output of our home regions, supplemented by trade in true ecological surpluses. In the process, we might well rediscover that meeting friends while bicycling home is more fun than spending lonely hours commuting on congested highways. The bioregional movement has compiled many inspiring examples on how to live "at home."[6]

Cities must simultaneously become more livable while increasing their density and becoming less auto- and resource-dependent. This requires phasing out the routine provision of physical and institutional infrastructure that imposes a resource-intensive lifestyle on generations to come. (The sprawling, inefficient urban form that accompanied the rise of the automobile in the 1950s and 1960s will be with us for many decades.) Suitable initiatives that are much discussed but less frequently delivered include: planning for high-density, high-amenity downtown restoration; promoting the use of renewable energy in commercial and housing developments; reallocating urban space, particularly road and other auto-oriented areas, to low-cost housing and public open space; imposing disincentives on auto use while creating incentives to encourage public transit, walking and bicycling; and using the tax system — rewards *and* penalties — to encourage urban development, urban land trusts, co-operative housing, etc., dedicated to sustainability principles. The point is that to keep cities within ecological limits, we must shift from planning ever-increasing capacity to making those limits felt and real through economic incentives and artificial bottlenecks in urban infrastructure.[7] This is a simple variation on the theme of managing demand rather than increasing supply.

Economic activity was once a means to an end; it was what people did to enable them to enjoy life. Today the economy is more an end in itself; both

Figure 4.6: How about learning to live in place? Becoming more regionally self-sufficient both increases the incentive to husband our own natural capital and reduces our dependence on other people's (temporary?) surpluses.

people and "the environment" are sacrificed to maintain the economy (or, more specifically, the existing wealth and power relationships within an increasingly global economic order). Indeed, the increasing concentration of economic power in the hands of fewer and fewer giant corporations and financial institutions with no commitment to place is excluding more and more people from effective participation in economic and political life. For ordinary citizens, it seems that globalization is creating a world of "powerless places at the mercy of placeless powers" (anon.).

Sustainability requires that we reclaim the economy in the service of people and their communities. The purpose of economic activity should be to foster material security where people live rather than to promote mindless consumption to maintain the world's financial centers at the expense of the ecosphere. It may seem paradoxical, but global security is likely to find its deepest roots in strengthened community and regional economies. No power on Earth can manage globally. However, if individual bioregions learn to live on the sustainable use of their own resources supplemented by ecologically balanced trade, the net effect would be global sustainability.[8]

Achieving the ideal will require establishing a balance between local and external control over regional resources and strengthening local management's hand over production and distribution. It would also require explicit policies to encourage local production for local consumption and a shift away from reliance on imported resources. In short, we need to restore values and incentives that encourage local populations to protect the long-term productivity of natural capital in their own home regions and to develop ecologically balanced trading relationships that do not compromise the livelihoods and prospects of people in other regions.

This is not an argument for decentralization of nations into totally autonomous regions but rather for a restoration of balance between the degree of centralization made possible by modern communications and transportation technology and the amount of local authority required: a) to manage human-ecosystem relationships in a manner that is sensitive to local conditions at the scale of ecosystems; b) to re-establish a sense of "connectedness" between human populations and the ecosystems that support them; and c) to reduce the alienation that intrudes between people and their employment when resources and capital are owned or controlled by absentee landlords. We also recognize that in any integrated world different functions of government require different distributions of authority between local and central agencies. A combination of local and centralized agencies and institutional arrangements seems in order. Certainly central (national or global) authorities with the necessary enforcement powers will be required to ensure that bioregions and nations are, in fact, thinking globally in their local actions and to co-ordinate the implementation of efforts to protect (or rehabilitate) the global commons.

All of this assumes, of course, that there is general agreement on the nature of the policies and actions required for global sustainability and the political will to follow them through. (Judging from the historical record, this last assumption may well be the weakest component in any global action plan.)

We are also aware (as Ecological Footprint analysis reveals) that today's large urban industrial regions pose an enormous challenge for any approach to sustainability based, even in part, on increased regional self-reliance. However, this is not so much an argument against the model as it is a warning that the current pattern of urbanization, with its extreme dependence on external flows, is inherently unsustainable. How dependable are those external flows? What is the likely impact of climate change on sources of supply? What would follow if producer regions are forced to reclaim surpluses to support their own populations? At the very least, in an era of increasing uncertainty and global change, dependent regions might consider formalizing their relationships with their suppliers to enhance the security of imports while striving both to increase local production and reduce demand. This would conform to the general bioregional vision in which local economies are scaled for compatibility with (trade-augmented) regional boundaries and both are adequate to support the corresponding regional population indefinitely.[9]

It should be clear from the above that the sustainability we are discussing involves a global steady-state, an economy of regions whose aggregate throughput of energy and resources has been stabilized with some safety margin below maximum carrying capacity (maximum sustainable load). There can be no further *material* growth. This does not mean, however, that all GDP growth must necessarily cease. Indeed, we noted earlier that growth is a pressing moral imperative for those whose needs are not being met, and industrialised countries have not yet found ways to maintain their standard of living without continued economic growth. One hopeful strategy to deal with this dilemma involves massive improvements in the efficiency of economic activity so that growth in consumption of goods and services is "decoupled" from growth in the use of energy and material. In theory, this should permit an *increase* in consumption to be accompanied by a *decrease* in resource use. In fact, this "dematerialization" of economic goods and services must proceed faster than economic growth to produce the necessary reduction in humanity's total load on the ecosphere. The political attractiveness of this approach is self-evident — it enables the rich to maintain their high material standards while freeing up the ecological space needed for the poor to increase theirs.

Ecological Footprint analysis supports the findings of numerous other studies that in the industrial countries a four- to ten-fold reduction in material and energy intensity per unit of economic output is required for global sustainability.[9] Some researchers refer to the desired state as the "factor-10" economy. Such an unprecedented efficiency gain represents a daunting technological goal but even this is not enough. As noted earlier in this chapter, enhanced efficiency must be accompanied by complementary policies that capture the anticipated economic gains to prevent their rebounding consumptively through the economy.

One essential complement is ecological tax reform. By raising prices closer to the full social cost of goods and services, substantial resource depletion taxes, marketable quotas on natural capital inputs, or similar taxes on resource consumption would not only encourage conservation but also: a) stimulate the search for the needed material- and energy-efficient manufacturing technologies; b) pre-empt any resultant cost savings, thereby preventing the economic benefits of efficiency gains from being redirected to additional or alternative forms of consumption; and c) generate an investment fund that could be used to rehabilitate important forms of self-producing natural capital. Meanwhile significant *reductions* in value-added, payroll and income taxes would reduce upward pressure on wages and salaries. Since both measures increase the attractiveness of labor relative to resources and capital, a positive side-effect of sustainability reform should be to increase the demand for labor.

There is more potential good news. Higher taxes and prices on energy and materials favor reuse, repair, reconditioning, and recycling, all of which are less material- and more labor-intensive than replacement manufacturing. Such product-life extension activities therefore tend to substitute small-scale, labor- and skill-intensive, locally integrated enterprises for large-scale, energy- and capital-intensive extractive and primary manufacturing industries. In summary, if managed properly, the net effect of the coming efficiency revolution should be not only less consumption and waste but also more employment opportunities and greater regional self-reliance.[10]

There are, of course, problems with any such revolutionary proposal, no matter how theoretically attractive it may appear. Public ignorance, irreducible scientific uncertainty, the power of vested interests, and the large potential costs associated with required structural adjustments to the economy all present barriers to the decisive political action required for the "factor-10" scenario. (How receptive would today's politically cynical electorate be to any suggestion of massive tax reform?) Thus, while the efficiency revolution promises a great deal, our social and political institutions may not be able to deliver the technological goods. In these circumstances, the mounting pressure of population growth, rising expectations and increasing competition might push ecological decline and social unrest to the point where the presently rich may be forced to accept lower material standards in exchange for enhanced ecological and geopolitical security.

To many this is the scary part of the sustainability conundrum — we are victims of a pervasive cultural myth that a steady-state (or diminished) economy equals deprivation. This is assuredly not the case. With some assistance from human ingenuity, the ecosphere can produce material adequacy for all. It would be foolhardy, however, to underestimate the difficulty associated with lowering the material expectations of today's consumer society. Both the perceived risk associated with staying our present course and the reward anticipated from changing our tack will have to be very great indeed! With luck, therefore, our brave new sustainable world will also be the kind of society that satisfies people's non-material needs. This has always been the great hollow void in contemporary political economy and filling it will require great personal and public investment in rehabilitating our social capital.

The optimists among us will greet this crisis-induced transition as humanity's last opportunity to become truly civilized and at home on planet Earth. Certainly recognition of the supra-economic role of natural capital is at least the first step toward a fuller eco-enlightenment. The transition will also require a shift of political emphasis from the quantitative to the qualitative dimensions of social evolution unprecedented in our techno-scientific age. We must now focus on how to improve human welfare by means other than sheer growth. Even those at the center of the sustainable development debate have tended to forget that growth simply means getting bigger while development means getting better. Having grown to the max, it is time that humanity began to concentrate on developing its full potential. (In theory, this shouldn't be too difficult — each of us does much the same thing in the course of our individual lives!)

There is increasing evidence that industrial society may, in fact, be prepared to abandon careless youth for more responsible maturity. For example, "The Natural Step" program in Sweden has developed principles to reduce the impacts of economic production on nature that are already widely used by Swedish local governments, industries and schools. Consistent with carrying capacity concepts, these principles maintain that human-produced substances and substances extracted from the earth must not be permitted to accumulate in the ecosphere, and that industry must avoid any manipulation of the ecosphere that diminishes its productivity or diversity.[11] A tiny glimmer to be sure, but enough to lend hope that there may soon be light enough to see our way clear to a sustainable future.

In summary, if the basic message of Ecological Footprint analysis is true, sustainable development is more than simple reform. As shown by even the preceding preliminary sketch, it will require a transformation of industrial society far beyond anything the political process has been willing to contemplate to date.[12] To those who say that any such vision is economically impractical and politically unrealistic we can only respond that the prevailing vision is ecologically destructive and morally bankrupt (to say nothing of potentially lethal). Surely what is politically realistic is determined by circumstances, and with global ecological decline the relevant circumstances have changed. The present challenge, then, is to raise the general level of awareness of this reality to the point where political consensus forms around the necessary policy initiatives. The alternative is to stay the present course until accelerating decline removes any lingering doubt that we are facing a global crisis. By then, of course, it will be too late to organize a reasoned, effective and globally co-ordinated response. Fortunately, this scenario may be losing ground — people are beginning to comprehend the ecological bottom-line: no ecosphere, no economy, no society (or, more simply for the business-minded: no planet, no profit).

Notes

1. Stefan Schmidheiny, *Changing Course* (Boston: MIT Press, 1992); Wolfgang Sachs, "The Gospel of Global Efficiency" (Nyon, Switzerland: IFDA Dossier 68, 1988) (quote is from p. 33); Donella Meadows *et al.*, *Limits to Growth* (NY: Universe Books, 1972); Lester Brown *et al.*, "From Growth to Sustainable Development," in *Population, Technology, and Lifestyle: The Transition to Sustainability*, Robert Goodland, Herman E. Daly and Salah El Serafy, eds. (NY: Island Press, 1991/1992); Bruce Hannon, "Energy Conservation and the Consumer," *Science* Vol.189 (1975): 95-102; Paul Samuelson and William Nordhaus, *Economics*, 12th edition (NY: McGraw-Hill, 1985); Harry Sanders, "The Khazzoom-Brooks Postulate and Neoclassical Growth," *The Energy Journal* Vol.13, No.4 (1992): 131-148; Charles A.S. Hall, Cutler J. Cleveland and Robert Kaufmann, *Energy and Resource Quality* (NY: John Wiley & Sons, 1986); Robert Kaufmann, "A Biological Analysis of Energy," *Ecological Economics* Vol.6, No.1 (1992): 35-56.

2. Manfred Max-Neef, "Human Scale Economics: The Challenges Ahead," in *The Living Economy*, Paul Ekins, ed. (NY: Routledge, 1986). Other approaches to quality of life include Ian Miles, *Social Indicators for Human Development* (London: Frances Pinter Publishers, 1985); or the Social Caring Capacity approach which is briefly described in UBC Task Force on Healthy and Sustainable Communities, "Tools for Sustainability: Iteration and Implementation," in *The Ecological Public Health: From Vision to Practice*, Cordia Chu and Rod Simpson, eds. (Centre for Health Promotion, University of Toronto, and Institute of Applied Environmental Research at Griffith University, Australia, 1994).

3. Vicki Robin, "A Declaration of Independence — from Overconsumption" (Seattle: The New Road Map Foundation, 1994). To learn more about tools to achieve financial independence and to shift to low-consumption, high-fulfilment lifestyles read Joe Dominguez and Vicki Robin, *Your Money or Your Life* (NY: Viking Penguin, 1992), or *Simplicity: Notes, Stories and Exercises for Developing Unimaginable Wealth*, by Mark A. Burch (Gabriola Island/Philadelphia: New Society Publishers, 1995).

4. Robert Ornstein and Paul Ehrlich, *New World, New Mind: Moving Toward Conscious Evolution* (NY: Doubleday, 1989); Garrett Hardin, "The Tragedy of the Commons," *Science* Vol.162 (1968): 1243-1248; Fikret Berkes, ed., *Common Property Resources: Ecology and Community Based Sustainable Development* (NY: Belhaven Press, 1989).

5. Odum, W. 1982, "Environmental degradation and the tyranny of small decisions," *BioScience* 32 (9): p. 728-729.

6. See for example *Home! A Bioregional Reader*, edited by Van Andruss et al. (Gabriola Island: New Society Publishers, 1990) and other volumes in *The New Catalyst's* Bioregional Series.
7. See Mark Roseland, *Toward Sustainable Communities* (National Round Table on the Environment and the Economy, Ottawa, 1992); Herbert Girardet, *The Gaia Atlas of Cities: New Directions for Sustainable Urban Living* (NY: Doubleday, 1993).
8. Contrast this with the present global development model, which assumes all regions can be net importers of carrying capacity (i.e., it assumes the entire planet can run an ecological deficit!). See William E. Rees & Mathis Wackernagel, "Ecological Footprints and Appropriated Carrying Capacity: Measuring the Natural Capital Requirements of the Human Economy," in *Investing in Natural Capital: The Ecological Economics Approach to Sustainability*, ed. A.M. Jansson, M. Hammer, C. Folke, and R. Costanza (Washington: Island Press, 1994).
9. *Fresenius Environmental Bulletin* (special edition on the "Material Intensity Per Unit Service" [MIPS] project of the Wuppertal Institute für Klima, Umwelt, und Energie in Wuppertal, Germany), Vol.2, No.8, 1993; Paul Hawken, *The Ecology of Commerce: A Declaration of Sustainability* (NY: Harper-Collins, 1993); Paul Ekins and Michael Jacobs, "Are Environmental Sustainability and Economic Growth Compatible?," in: Energy-Environment-Economy Modelling Discussion Paper No. 7 (Cambridge, UK: Department of Applied Economics, University of Cambridge, 1994); John Young and Aaron Sachs, *The Next Efficiency Revolution: Creating a Sustainable Materials Economy*, Worldwatch paper 121 (Washington: The Worldwatch Institute, 1994); BCSD, *Getting Eco-Efficient*, report of the BCSD First Antwerp Eco-Efficiency Workshop, November 1993 (Geneva: Business Council for Sustainable Development).
10. William E. Rees, "Sustainability, Growth, and Employment: Toward an Ecologically Sustainable, Economically Secure, and Socially Satisfying Future," paper prepared for the IISD Employment and Sustainable Development Project (Winnipeg: International Institute for Sustainable Development, 1994). Revised version forthcoming in *Alternatives*, September 1995.
11. John Holmberg, Karl-Henrik Robèrt and Karl-Erik Eriksson, "Socio-Ecological Principles for a Sustainable Society," paper presented at Third Conference of the International Society for Ecological Economics (Costa Rica, 1994).
12. William E. Rees, "Achieving Sustainability: Reform or Transformation?" *Journal of Planning Literature*, Vol. 9, No.4, p. 343-361.

5

AVOIDING OVERSHOOT: A SUMMARY

We have shown that current human consumption of agricultural products, wood fiber and fossil fuel have an Ecological Footprint that exceeds available ecologically productive land by close to 30 percent. In other words, we would need an Earth 30 percent larger (or more ecologically productive) to accommodate present consumption without depleting corresponding ecosystems. To look at these data in an even more telling way, United Nations statistics show that the 20 percent of the world's population that lives in wealthy countries consumes up to 80 percent of the world's resources. This translates into the developed world *alone* occupying an Ecological Footprint larger than global carrying capacity:

   80% [of the world's resource consumption] of
   130% [humanity's Footprint as compared to the global carrying capacity]
   = 0.8 x 1.3 = 1.04 (104%) [industrial countries' Footprint as compared to global carrying capacity].

In short, there's nothing left into which the rest of the world can grow (without eroding global life-support)! Our present ecological overshoot is indicated by the global degradation of forests, soil, water systems, fisheries and biological diversity.

Such trends demonstrate not only the overwhelming ecological impact of the present generation, but also the responsibility for change that rests with wealthy countries and the challenge that may confront future generations. Despite the best efforts of technology, the fact is that for the foreseeable future human well-being and security depend on the capacity of remaining stocks of natural capital to provide adequate flows of essential goods and life-support services. Human-made systems cannot substitute for the life-support functions of the ecosphere.

By showing the link between various competing human uses of nature and available ecological space, Ecological Footprint analysis provides a framework for visualizing and communicating the phenomenon of "overshoot." While the notion of "limits to growth" may suggest discrete boundaries and hard edges, the expanding economy does not crash into ecological limits as a car would crash into a wall. Natural limits are fuzzy and can be temporarily exceeded at the cost of drawing down nature's well. As a sustainable rate of withdrawal is crossed, no explicit warning light flashes the alarm — the quiet loss of natural capital is the only indication that total human load has exceeded carrying capacity. Often, even this degradation can be difficult to detect because the differences between ecosystems that are being used sustainably and those that are being degraded are subtle. (It may take years for rich but eroding agricultural land to suffer falling productivity.) Ecological Footprints provide a necessary "warning light" by revealing the stark disparities between demand and long-term resource availability. People seem especially to appreciate its use in helping to link the global ecological challenge to local consumption and local — even personal — decision-making responsibility.

EF analysis can therefore assist in the development of appropriate policy responses in a wide range of contexts from technology, policy and environmental assessment, through local, regional, and national planning, to the design of international treaties. Translating the ecological aspects of sustainability into a concrete common yardstick seems to bridge misunderstandings among groups with conflicting politics and differing interpretations of sustainability. International interest in the tool and the growing list of applications attest to the Ecological Footprint's analytic merit and its teaching value in communicating the sustainability imperative. Indeed, we feel the greatest strength of EF analysis is in creating public awareness and consensus around sustainability issues at both the local and global levels.

Creating Public Awareness

Effective political action requires solid public support. However, there is little evidence that many of "the public" understand the nature of global ecological change or appreciate the potential consequences of failing to respond. Public opinion polls tell us that most people are concerned about environmental problems, but few understand or accept the full implications of making the shift to a more sustainable society. This is unfortunate in an era when political "leadership" seems predicated on which way the winds of public opinion are blowing. At the same time it shows the importance of creating sound public understanding around sustainability issues.

It seems that general ignorance combined with immediate economic and political constraints at all levels of decision-making force us to make what appear to be unsustainable policy choices. However, rather than bemoan the constraints of "real-life" decision-making, which result in slow but incremental ecological destruction, non-government organizations (NGOs), planners, and policy analysts now have tools that allow them to estimate and reveal publicly the extent to which "development" decisions compromise a sustainable future.

For example, EF analysis makes clear that economic growth that is politically attractive has hidden and potentially greater long-term costs and places an extra burden of accountability on politicians. In a world at its carrying capacity, decision-makers have an obligation to approve only those technologies, development projects and growth strategies that *reduce* society's Ecological Footprint. Any option that does otherwise contributes to long-run instability and uncertainty in a negative-sum game that is detrimental to everyone. In particular, growth or development decisions that increase one group's overconsumption at the further expense of common-pool resources impose hidden costs on everyone as long-term losses outweigh the gains.

Sometimes the argument is put forward that economic growth and expansion in aggregate consumption is inevitable, and that every "development" option must be seized "otherwise somebody else will take it." This position, based on the vulnerability of common-pool resources, is not defensible in a world already showing signs of overshoot — not every country can simultaneously maximize its gains in a finite world. Ecological Footprint analysis can help all participants in the decision process, from active citizens through NGOs to government policy analysts, to monitor the cumulative impacts of conventional economic development. It thereby also provides the empirical basis for establishing the local regulatory regimes and international accords needed to protect common-pool assets and global life-support for the benefit of all.

Politicians are faced with immediate pressures to accommodate development proposals that seem fiscally attractive today. But if they — and their constituents — know also that such development will expand the community's Ecological Footprint at the expense of society's long-term interests then the decision process may be more sensitively balanced. Certainly EF analysis can help citizens hold decision-makers more accountable for actions ostensibly taken on their behalf. There is no longer any reason why society should passively accommodate unsustainable development proposals. The technological and regulatory means are available so that necessary development can be made into virtually zero-impact development. The warning signals are clear that without these adjustments so-called "inevitable growth" today means "inevitable destruction" tomorrow.

Today's superficial policy responses to the sustainability crisis simply feed into societal denial of the problem. Taking minimal action makes people feel as if they are doing something while not significantly changing their lifestyles. As one municipal planner has observed, "...there is a certain sexiness about an expensive lifestyle that [the perception of] going without just does not have...." Perhaps to free us from this pathology, we need spear groups at various political levels with enough conviction and self-confidence to accept the sustainability challenge and to resist the shallow attractions associated with further growth of the consumer society. In this context, the Ecological Footprint provides an interactive research tool to assist people to visualize the conflicts and challenge the assumptions of palliative approaches. EF analysis can help to probe the structure of cultural denial and expose the barriers to sustainability be they misunderstanding; inadequate information; incomplete or distorted world views; conflicting value systems; simple carelessness; economic desperation; fear of the future; or external sociopolitical constraints such as an increasingly competitive economic environment. More important, it can help show the relative *attractiveness* of an ecologically sound approach to sustainability compared to the expansionist model.

This latter point is the key to public support for sustainability initiatives. Holding to our present development course may well precipitate the human equivalent of the "boiled frog" syndrome as bit by unnoticeable bit the ecosphere deteriorates beyond recovery. By contrast, EF analysis can assist planners and policy analysts to recognize when we have reached the cumulative limits of incremental decisions while there is still time to avoid a serious crash. Raising to popular consciousness a sense that "the world is full" is prerequisite to acceptance of the policy adjustments that will be necessary to put us on a sustainable development path. In short, EF analysis can play a vital role in moving society to the point where people and their institutions accept the need to reorganize to secure human welfare and community livability in the face of a shrinking resource base.

Global change reminds us that the economy is embedded in the ecosphere, and human life is dependent on the maintenance of ecological life-support. The real challenge for community groups and policy analysts, then, is to gain acceptance for the idea that these realities impose real constraints on the development process. The fact is that consumption is limited by nature's reproductive capacity — over-consumption today means less natural capital and lower natural income tomorrow. This, in turn, may force future generations to accelerate the downward spiral as they erode remaining stocks of natural capital to meet their own consumption needs. In other words, life on Earth (including human life) can be sustained only within the limits of the dividends nature pays on our remaining stocks and future investments in natural capital. EF analysis provides a valuable accounting tool to track our investments and ensure that our real (ecological) wealth is adequate to support the growing human family.

To reiterate, our message is simple enough: sustainability requires that the human enterprise remain within global carrying capacity. If global carrying capacity is already overshot, and advanced countries have taken more than their fair share of the Earth's bounty, then these countries must find ways to reduce material consumption while maintaining their livability. Of course, even within rich countries, consumption is inequitably distributed so that even as we strive to reduce aggregate resource use, consideration must be given to improving the lot of those whose basic requirements are not met.

We must also recognize that people will tolerate measures to reduce material throughput only if they feel that such measures will provide a better future for themselves and their children than the next best alternative. Sustainability-as-sacrifice will not sell (and if survival is at stake, even the affluent need a

Figure 5.1: Reducing our Footprint.
Will my decision or activity contribute to that goal?

functional ecosphere)! Policy analysts and decision-makers must therefore be able to demonstrate to the public that improving our quality of life is still possible even as we reduce our Ecological Footprints. When testing a technology, project, program or policy for its sustainability merits, two questions must be asked:

- Will this decision or activity reduce people's Ecological Footprint?; and
- Will this decision or activity improve our quality of life?

Only those decisions or activities that satisfy at least one of these criteria without violating the other can move us toward sustainability. Since judgments about the relationships among consumption, quality of life, and sustainability are somewhat subjective, effective means should be established to ensure that citizens are able to participate in the overall planning process.

Developing Sustainability — Locally and Globally

Developing sustainability requires rethinking the way we govern and make decisions. We must question the way we organize our cities, plan our infrastructure, and live our lives, and we must begin the process now. The "wait and see" option preferred by the political mainstream is not a viable alternative. The Earth's biophysical systems are large, complex, self-organizing entities. This means there is typically a long lag time between economic cause and ecological effect. (For example, whatever global warming we may already have experienced is not the result of today's levels of greenhouse gases but rather the levels reached perhaps 40 years ago; even though CFC production may be winding down, ozone depletion may worsen for a decade and it may be a half century or more before stratospheric ozone returns to normal.) Thus, the temptation to wait until we are certain that a particular trend is fatal, dangerous or simply uneconomic before deciding on corrective action leads us into an ecological trap. At best, the delay simply further entrenches our unsustainable lifestyles, making change the more difficult; at worst, it will be too late to do anything to reverse the trend. The longer we stall the less likely it will be that learning to live within carrying capacity can be combined with maintaining or improving our quality of life. Indeed, we will have lost the opportunity for a smooth transition to sustainability, putting society at increased risk of ecological instability and sociopolitical chaos.

Unquestionably, Ecological Footprint analysis and related studies challenge the fundamental assumptions driving economic globalization, international development, and population policies. If humanity is already at its ecological limits we must recognize that excessive material and energy consumption by one group compromises the opportunities for consumption by others, present or future, because of natural capital depletion. Nevertheless, high-income countries still do not encourage population reductions or lower consumption, but are more worried about the aging of society and slack economic growth. Indeed, today we encourage more consumption by the wealthy as a means of increasing the incomes of the poor through trade. Certainly the impoverished quarter of humanity should be able to increase its standard of living — everyone has the right to a decent life and material sufficiency — but is inflating the global economic balloon even further the best way to achieve this? Any increase in aggregate economic production based on additional erosion of natural capital will increase humanity's ecological deficit, contributing to an accumulated debt which, like a money debt, must someday be paid off. However, ecological overspending is even more pernicious than a fiscal deficit (like those which many governments are currently also running up) because beyond a certain point depleted species and essential processes may not recover and cannot be replaced.

The pursuit of environmentally-sound technologies cannot be used as an excuse to avoid questions of over-consumption and increasing material inequity. On the one hand, any comprehensive strategy for sustainability must protect those assets upon which all future generations depend for survival. On the other hand, it must address the ethical dilemma posed by our failure *even in overshoot* to meet the basic needs of a quarter of the present generation. This inequity has both domestic and international dimensions. Ecological Footprint analysis reveals that most high-income countries cannot support themselves on local carrying capacity alone (Canada and Australia may be lucky exceptions with their relatively small populations and extensive land bases). Maintaining our present consumer lifestyles with present technology not only draws down our domestic resource wells but also depends on the continuous appropriation of carrying capacity from the global common pool and on imports from low-income countries. New rules are needed to ensure fairer access by everyone to sustainable flows of goods and services from the world's ecosystems.

Indeed, under existing terms of trade and rules for economic exchange, the resource hunger of the rich threatens to liquidate the world's ecological assets — the very basis of life. Globalization of the economy may provide a few more resource-rich years and a new round of seemingly glorious growth but in the present circumstances this is at the cost of long-term productivity and survivability. Global markets give rampant global demand access to the world's last remaining pockets of unexploited or "under-utilized" natural capital, inevitably accelerating their depletion. Malaysia's forests disappear to satisfy Japanese timber hunger; Russia "develops" by opening its forest and fossil fuel stocks to Western exploitation; Canada's fisheries collapse from over-exploitation to satisfy global demand; the Southern Hemisphere suffers an ozone hole caused mostly by northern technologies.

The point is that as living standards rise, more and more people live on ecological carrying capacity "imported" from somewhere else. The obvious follow-up question is: how long will it be before we run out of "somewhere else"? (Answer: we already have.) If the so-called "advanced" countries continue to promote a lifestyle whose satisfaction will require the equivalent of several more planets, they are, in effect, blindly planning their own demise. The greatest contribution the developed world can make to sustainability is to reduce its resource consumption by all means at its disposal. The "factor-10" efficiency revolution (see chapter 4) may be the politically most acceptable approach, but there may well be greater ecological, community and personal merit in learning to live more simply so others can live at all.

There is also a practical consideration — growing populations, increasing material expectations and a deteriorating resource base can only exacerbate social and international conflicts, decreasing everyone's quality of life. Ecological Footprint calculations make clear that every decision resulting in the appropriation of more resources by those who already consume more than their fair share is a conscious choice against ecological, social and economic sustainability. Thus, while sustainability may constrain the economic choices of the affluent today, it will keep more options open for everyone in the future — particularly the option of not suffering ecological deterioration and geopolitical insecurity.

In a world at the limit of its carrying capacity, with growing populations and rising material expectations, how to provide adequately and fairly for everyone remains the major challenge. We think the Ecological Footprint tool can help to meet this challenge; it raises essential questions about long-term sustainability ignored by other approaches; it facilitates the comparison of policy choices; and it could be used to monitor progress towards reducing the sustainability gap.

Of course, there remains much scope for improving the tool and developing further applications. As previously emphasized, there is no shortage of sustainability strategies. What we lack is intellectual and emotional acceptance of the fact that humanity is materially dependent on nature and that nature's productive capacity is limited. It is here that Ecological Footprint analysis comes into its own. Its major strength is its ability to communicate biophysical realities simply and clearly and thus contribute to the needed shift in social consciousness. Practical examples both raise additional cautionary flags to which various interests can react and provide concrete guidelines for the development of suitable policy responses. Over time, EF analysis can contribute to the development of a broadly-based program of reforms to move us in the direction we all say we want to go.

A final reminder: Ecological Footprint analysis is not about how bad things are. It is just about *how* they are — and what we can do about it. Our analyses do suggest that society will have to make significant adaptations in the transition to sustainability. However, to the extent that the assumptions and prescriptions of the ecological approach are a better reflection of material reality than those of mainstream models, our conclusions are a good news story. The bad news is that most of the world seems committed as never before to the well-worn expansionist path — a habit that we now need to break.

GLOSSARY

Acre see hectare.

Appropriated signifies captured, claimed, co-opted, or made use of exclusively for oneself, sometimes without permission. When appropriating carrying capacity, one uses biological productivity from somewhere else, legitimized by mere purchasing power. Appropriations from the global commons require no payment.

Aquatic and marine ecosystems or water-based ecosystems represent all freshwater systems and oceans.

Biodiversity, according to the World Conservation Union (IUCN), is "the variety of life in all its forms, levels and combinations. Includes ecosystem diversity, species diversity, and genetic diversity."

Biological productivity refers to nature's capability to reproduce and regenerate, thereby accumulating biomass. Biological productivity of a given land category is determined by dividing the total biological production by the total land area available in this category.

Biologically productive land is land that is sufficiently fertile to accommodate forests or agriculture, i.e., there is significant net primary production.

Biomass or biomatter is the amount of living organic matter of an ecosystem — usually measured in dry weight.

Biophysical refers to the living and non-living components and processes of the ecosphere. Biophysical measurements of nature quantify the ecosphere in physical units such as cubic metres, kilograms or joules rather than in dollars.

Carrying capacity is conventionally defined as the maximum population size of a given species that an area can support without reducing its ability to support the same species in the future. In the human context, William Catton defines it as the maximum "load" (population x *per capita* impact) that can safely and persistently be imposed on the environment by people.

Consumption refers to all the goods and services used by households. This includes purchased commodities at the household level (such as clothing, food and utilities), the goods and services paid for by government (such as defence, education, social services and health care), and the resources consumed by businesses to increase their assets (such as business equipment and housing).

Ecological Footprint is the land (and water) area that would be required to support a defined human population and material standard indefinitely.

Embodied energy of a commodity is the energy that is used during the entire life cycle of the commodity for manufacturing, transporting and disposing of the commodity.

Energy-to-land ratio the amount of energy that can be produced per hectare of ecologically productive land. The units used are gigajoules per hectare and year, or $GJ/ha/yr$. For fossil fuel (calculated as CO_2 assimilation), the ratio is 100 $GJ/ha/yr$.

Erosion is the process of soil and nutrient loss, which leads to a decline in biological productivity. Can also be used metaphorically to refer to depletion (e.g. of natural capital).

Giga- means one billion. For example, one gigajoule (or GJ) refers to one billion joules.

Global economy refers to the emerging international economy characterized by free trade in goods and services, unrestricted capital flows and weakened national powers to control domestic economies. In 1990, world trade flows amounted to $4.3 trillion US compared to the gross global product of approximately $20 trillion. This indicates that national economies are becoming one tightly linked global economy.

Hectare (or ha) spans 10,000 square metres, the equivalent of a 100 metre square. There are 2.47 acres in one hectare.

Hydrological cycle the natural cycle of water from evaporation, transportation in the atmosphere, condensation (rain), and flow back to the ocean.

Joule (or J) is the physical measurement for work. One joule corresponds to the work of lifting one kilogram ten centimetres off the ground. It can also be used to measure heat energy. One kilocalorie, an older energy unit, corresponds to 4.1868 kilojoules.

Kilo- stands for thousand. Hence a kilogram (or kg) is 1,000 grams, or a kilojoule is 1,000 joules.

Life-support systems, according to the World Conservation Union (IUCN), refer to the biophysical processes "that sustain the productivity, adaptability and capacity for renewal of lands, waters, and/or the biosphere as a whole."

Low entropy energy refers to high quality energy, or energy that is concentrated and "available." For example, electricity is considered to be the energy carrier with the lowest entropy (i.e., highest quality) as it can be transformed into mechanical energy at efficiency rates well above 90 percent. In contrast, fossil fuel's chemical energy can only be converted into mechanical energy at a typical efficiency rate of 25 (cars) to 50 percent (modern power stations). The chemical energy of biomass is of lower quality.

Mega- means million. For example, one megajoule (or MJ) is equal to one million joules. One litre of gasoline contains about 35 megajoules of energy.

Net Primary Production is the energy or biomass content of plant material that has accumulated in an ecosystem over a period of time through photosynthesis. It is the amount of energy left after subtracting the respiration of primary producers (mostly plants) from the total amount of energy (mostly solar) that is fixed biologically.

Overshoot, according to William Catton, is "growth beyond an area's carrying capacity, leading to crash."

Peta- means one quadrillion (a thousand trillions, or one with 15 zeros). PJ is the abbreviation for petajoule.

Photosynthesis is the biological process in chlorophyll-containing cells that transforms sunlight into plant matter (or biomass).

Resource stock comprises the resource in its entirety, while the **resource flow** describes the amount of resources harvested *per unit time*. To harvest a resource stock sustainably, the harvest must remain smaller than the net production of the stock. Stocks are measured in mass, volume or energy; and flows in mass, volume or energy *per unit of time*. (See also "Watt".)

Square kilometre (or km^2) is a 1,000 metre square, and contains 1,000,000 square metres, 100 hectares or 247 acres.

Sustainability gap refers to the difference between ecological production and current human over-consumption. Developing sustainability means reducing the sustainability gap.

Tera- refers to one trillion (or one thousand billions). For example, one terawatt (or TW) refers to one trillion watts. The solar energy that shines onto Earth amounts to 175,000 terawatts.

Trade balance refers to the net trade flow of a country (exports minus imports), usually measured in monetary units. However, this could also be calculated in terms of embodied Footprint, leading to an understanding of the net drain or gain in appropriated ecological productivity.

Watt (or W) is the physical measurement unit for power. One watt corresponds to lifting one kilogram one metre every ten seconds, or to one joule per second. Energy would correspond to a "stock," and energy per time (or power) to a "flow."

For further information on ecological footprint analysis, or to
share your own footprint work, please write to:

Mathis Wackernagel
Centro de Estudios para la Sustentabilidad
Universidad Anahuac de Xalapa
Apt Postal 653
91180 Xalapa, Ver.
MEXICO

and

William Rees
School of Community & Regional Planning
University of British Columbia
6333 Memorial Road
Vancouver, BC V6T 1Z2
CANADA

New Society Publishers focuses much of its publishing program upon the developing of sustainability in a number of different spheres. If you have enjoyed this book, you may want to check out the following titles also:

- *Simplicity: Notes, Stories and Exercises for Developing Unimaginable Wealth*, by **Mark A. Burch.** Mark eloquently explores voluntary simplicity as a means to personal sustainability, sketching a practical, enriching alternative to the culture of consumption, as well as paths you can take to joyously change your life.
  6" x 9". 144 pages.
  Canada Pb: $15.95. ISBN: 1-55092-269-6
  USA Pb: $12.95. ISBN: 0-86571-323-5

- *Whole Life Economics: Revaluing Daily Life*, by **Barbara Brandt.** Charting an emerging economics of empowerment, this book is a practical guide to how the economy could work for both people and planet.
  6" x 9". 240 pages.
  Canada Pb: $18.95. ISBN: 1-55092-209-2
  USA Pb: $14.95. ISBN: 0-86571-266-2

- *Whose Common Future? Reclaiming the Commons*, by **The Ecologist Magazine.** Tracing the world's environmental crisis to the dismantling of the commons world-wide, *The Ecologist Magazine* editorial team analyses the forces behind the destruction of commons regimes, and highlights effective strategies for protecting and recovering them.
  5½" x 8¾". 224 pages.
  Canada Pb: $17.95. ISBN: 1-55092-221-1
  USA Pb: 14.95. ISBN: 0-86571-277-8

- *Home! A Bioregional Reader,* **edited by Van Andruss, Christopher Plant, Judith Plant, and Eleanor Wright.** More than 40 contributors make this a compelling introduction to the classic thought and literature of bioregionalism — the exciting movement for (re-)creating an ecological, sustainable society rooted in community and a culture of place.
  8½" x 11". 192 pages.
  Canada Pb: $19.95. ISBN: 1-55092-007-3
  USA Pb: $16.95. ISBN: 0-86571-188-7

www.newsociety.com